ROUTLEDGE LIBRARY EDITIONS:
POLITICAL GEOGRAPHY

T0251091

Volume 5

THE GEOGRAPHY OF
ENGLISH POLITICS

THE GEOGRAPHY OF ENGLISH POLITICS

The 1983 General Election

R.J. JOHNSTON

Routledge
Taylor & Francis Group

LONDON AND NEW YORK

First published in 1985

This edition first published in 2015
by Routledge
2 Park Square, Milton Park, Abingdon, Oxon, OX14 4RN

and by Routledge
711 Third Avenue, New York, NY 10017

Routledge is an imprint of the Taylor & Francis Group, an informa business

British Library Cataloguing in Publication Data
A catalogue record for this book is available from the British Library

ISBN: 978-1-138-80830-0 (Set)
eISBN: 978-1-315-74725-5 (Set)
ISBN: 978-1-138-80149-3 (Volume 5)
eISBN: 978-1-315-75033-0 (Volume 5)
Pb ISBN: 978-1-138-80893-5 (Volume 5)

Publisher's Note
The publisher has gone to great lengths to ensure the quality of this reprint but points out that some imperfections in the original copies may be apparent.

Disclaimer
The publisher has made every effort to trace copyright holders and would welcome correspondence from those they have been unable to trace.

Printed and bound by CPI Group (UK) Ltd, Croydon, CR0 4YY

The Geography of English Politics

The 1983 General Election

R.J. Johnston

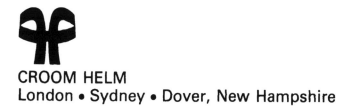

CROOM HELM
London • Sydney • Dover, New Hampshire

© 1985 R.J. Johnston
Croom Helm Ltd, Provident House, Burrell Row,
Beckenham, Kent BR3 1AT
Croom Helm Australia Pty Ltd, Suite 4, 6th Floor,
64-76 Kippex Street, Surry Hills, NSW 2010, Australia

British Library Cataloguing in Publication Data

Johnston, R.J.
   The geography of English politics: the 1983
   general election.—(Croom Helm series in
   geography and environment)
   1. Great Britain, *Parliament*—Elections, 1983
   2. Voting—Great Britain—History—20th century
   I. Title
   324.941'0858   JN956

ISBN 0-7099-1441-5

Croom Helm, 51 Washington Street, Dover,
New Hampshire, 03820, USA

Library of Congress Cataloging in Publication Data

Johnston, R.J. (Ronald John)
   The geography of English politics.
   (Croom Helm series in geography and environment)
   Bibliography: p.
   Includes index.
   1. Great Britain. Parliament—Elections, 1983.
2. Elections—Great Britain. 3. Voting—Great Britain.
4. Social classes—Great Britain—Political activity.
I. Title. II. Series.
JN956.J65 1985   324.941'0858   84-27512

ISBN 0-7099-1441-5

Typeset by Columns of Reading
Printed and bound in Great Britain by
Biddles Ltd, Guildford and King's Lynn

CONTENTS

FIGURES

TABLES

PREFACE

Each general election in Britain provides a new data set to be analysed and reanalysed. The 1983 election will be no different; indeed, it is likely to produce an even greater spate of studies than usual, because of the new element – the first election fought by the Liberal Party in alliance with the new Social Democrat Party. At the time when this book was completed (early 1984), four books had been advertised (Butler and Kavanagh, 1984; Dunleavy and Husbands, 1984; McAllister and Rose, 1984; Waller, 1984b) although none was available.

Why, then, produce this book? The major reason is supported by a review (Berrington, 1984), published after this book was completed, of the most recent book-length study of British voting behaviour – *Decade of Dealignment* (Sarlvik and Crewe, 1983). Berrington notes that at the end of the period covered in the book

> Mrs. Thatcher's victory in 1979 . . . saw an accentuation of the geographic division of Britain into two nations – with Southern England, the Midlands and Wales swinging sharply to the right, Northern England moving much more softly and Scotland virtually marking time (p. 117).

But Sarlvik and Crewe ignore this changing geography. As Berrington puts it

> The biggest gap lies in the failure to explore the relationship between geography, class and political affiliation (p. 119).

The present book is aimed directly at that gap, providing an exploration of geographical variations in the relationship between class and voting, in England, on 9 June 1983.

The need for a geographical exploration of voting in England is the basic rationale for the book. Allied to this is a perceived need to develop the explanations for geographical variations in voting behaviour. The commonly-presented 'explanations', based on notions of contagious effects, are unconvincing, and so, in addition to providing analyses of 'who voted what, where', attempts are made here to provide a better understanding of what is described. The descriptions themselves are not straightforward, however, since electoral data do not report on who voted what, only how many votes each party received. This problem is tackled here using a method of estimating 'who voted what, where' through a combination of electoral, census and survey data.

The principal goal of this book is to demonstrate the need for a geographical perspective in the study of English voting behaviour. This is not to claim that others have not perceived that need previously (the works of Miller and of Crewe are important in this respect). But there is a tendency to treat England as an homogeneous place, a tendency which this book surely destroys. Demonstrating the need for a geographic perspective calls for developments in data handling, and the book also argues for a particular approach, claiming that it provides better information about the geography of voting than do previous studies. Finally, demonstration – however sophisticated – is of little use without explanation, and the third goal of the book is a development of geographical understanding.

The core of the book is a series of statistical analyses, using regression models to describe and account for the estimated values of 'who voted what, where'. Those estimates take the traditional class-basis of English electoral behaviour as given; they show the extent of variability in the partisan choices of people in similar socio-economic positions (the usual shorthand term 'class' is used throughout the book, with the adjectives 'middle' and 'working' referring to white-collar and manual occupational groups respectively). The analyses (Chapters 5-9) are set within a general framework which provides an explanatory account for the patterns revealed (Chapters 1, 4 and 10), whereas two others (2 and 3) indicate the salience of these patterns for the overall election result – i.e. the allocation of seats among the Conservative, Labour and Alliance contestants. The statistical procedures used are straightforward and well known. Their use was valid (see Appendix 4) and they were in general very successful. More sophisticated modelling could add to the results presented here, but would only enhance the basic points made regarding the geography of voting.

In producing these analyses within nine months of the election, I am indebted to a large number of people. The staff of the SSRC Data Archive have been most helpful in providing tapes containing the 1981 Census data, by parliamentary constituency, and the BBC/Gallup Election Survey; Jeff Martin, Paul Leman, Ian Dukinfield and Wilf Webster in the University of Sheffield's Computing Services gave much valuable assistance in mounting these tapes and extracting the needed data. A grant from the University of Sheffield Research Fund covered the costs of some of the coding, and I am grateful to Barbara Woods for her efficiency at this. In producing the book, I am deeply indebted once again to Joan Dunn for all her secretarial work, to Graham Allsopp and Paul Coles for their cartographic achievements, and to Peter Morley, John Owen and David Maddison for excellent photographic assistance. Mike Bradford as editor encouraged me to produce the book and made some valuable editorial comments.

More generally, my work in this field has been encouraged and assisted by two colleagues, who have no responsibility for what I have done but without whose collaboration I could not have developed the ideas presented here. Alan Hay's recognition of the technical solutions to the data problems I had identified made the format of the analyses possible, while his scepticism concerning my application of those solutions restrained me to some extent; I am extremely grateful to him for his willingness to become involved in this work and for his characteristic generosity towards fellow-researchers. Finally, I am deeply indebted to Pete Taylor, in whose intellectual company I have learned much over the past decade; his insights into the political and the technical aspects of studying the geography of elections have been of immeasurable value, and his abilities to rethink a wide variety of issues have provided a continuing and exciting stimulus. Whether or not these two appreciate what I have done with the resources they have helped me to develop, my appreciation of their contributions cannot go unsaid.

Finally, my thanks are due to my wife Rita and my children, Chris and Lucy, who now seem to accept that every general election means hours at computer and desk.

1 CLASS AND VOTING BEHAVIOUR

> Class is so much the most important conditioner of political allegiance that it becomes difficult to disentangle it from others (Pulzer, 1967, p. 107)

The statement by Pulzer is typical of a large majority of the analyses of voting behaviour in Britain. It does not claim that class – basically the division of society into white-collar (or middle-class) and blue-collar (working-class) sectors – is the sole influence on voting behaviour. But it does clearly state that class is the major determinant, with the implication that others – age, sex, race, religion, etc. – are residual influences, to be used to account for minor deviations from the general pattern. Others agree, leading Butler and Stokes (1974, p. 77) to the conclusion that

> Our findings on the strength of the links between class and partisanship in Britain echo broadly those of every other opinion poll or voting study.

The basis of this assumed correlation lies in the origins and nature of the two political parties that dominated in Britain during the period (1945-80) in which most of the studies reviewed here were conducted. The Labour Party developed to represent the interests of the working class against those of capitalist employers, has enduring links with the trade union movement, and has a constitutional position dedicated to the replacement of the capitalist mode of production by socialism. Not surprisingly, therefore, its main constituency has been the working class although it gains substantial support in certain parts of the middle class and most of its national leaders and MPs have had middle-class occupations, if not family backgrounds. The Conservative Party, in contrast, exists to advance the interests of those who benefit most from the capitalist mode of production, though its nationalist ideology presents it as representing a system that is the best for all and its links to other issues – law and order, defence, empire – win it considerable support outside the middle class.

Analyses of voting behaviour have generally confirmed this stereotype portrayal of the class basis of the two main British parties. These analyses are of two main types. The first are survey-based, involving questionnaires administered to samples of the electorate to elicit details of their voting behaviour, political beliefs, social, economic and political attitudes, and personal circumstances. The second are aggregate data analyses, in which voting patterns by places (usually parliamentary constituencies) are compared statistically to the population characteristics in those places. Both types are relatively recent innovations to academic work on voting. Large-scale *academic* surveys of voters (as against opinion polls) have only been conducted in Britain since about 1960; detailed statistical analyses of constituency data have only been possible since the late 1960s, with the publication of data from the 1966 sample Census of Population at the constituency scale.

The implications of the literature on voting behaviour in Britain are that class dominates electoral choice, and that there are no other major cleavage lines within British society. (In recent decades, the emergence of separate political parties in Northern Ireland and the growth of the nationalist parties – SNP and Plaid Cymru – in Scotland and Wales confine the validity of this statement to England, hence the concentration on that country alone in this book: for treatments of the other countries see, *inter alia*, Miller, 1981 and Balsom *et al.*, 1982.) Thus, for example, Butler and Stokes (1974, p. 120) wrote that

> Certainly there are variations in British electoral behaviour which cannot be attributed to class in any simple way . . . [Discussion of these] will not lead us away from class so much as towards an understanding of additional ways in which social structure is reflected in electoral behaviour.

This careful statement gives primacy to class, but indicates the need to study the residual features. Other analysts have not been so careful. In particular, spatial variations in voting behaviour have been underplayed by many commentators, leading to a 'one nation' view of English elections:

Electoral behaviour came to display a considerable degree of geographical homogeneity since an elector in Cornwall would tend to vote the same way as an elector from a similar class in Glasgow regardless of national or locational differences (Bogdanor, 1983, p. 53).

The primary purpose of the present book is to refute such statements, and to argue that *where* a voter lives is a considerable, and very relevant, influence on electoral choice.

Some other analysts (notably Miller, 1977, 1978) have also advanced this case, though their data base is rarely strong. Others have argued against the 'dominant class-cleavage' position, but have not introduced geographical variables.

The remainder of this chapter reviews the literature on electoral choice in England, with particular reference to the 'one nation' position. This is not done to set up a 'straw man' which can then be knocked down in subsequent chapters, but rather to provide a background résumé of the general beliefs about English electoral behaviour.

The Dominance of the Class-Cleavage

Many societies are divided politically in several ways, producing a variety of cross-cutting cleavages. (The best introduction to these is the seminal work of Stein Rokkan: see Rokkan, 1970; for a brief summary, see Taylor and Johnston, 1979, ch. 3.) In England, on the other hand, it has traditionally been deduced from analyses of survey and electoral data that only one cleavage – that based on class – is relevant.

Origins

One of the first clear, social scientific statements of the dominance of a class-cleavage in British politics was provided by Robert Alford, an American political scientist. His, still-cited, book *Party and Society* (1963) reported a comparative study of survey data from four countries – Australia, Canada, Great Britain, and the United States – which produced the conclusion that the class-cleavage was greatest in Great Britain. He quoted the conclusions of Bonham (1954, p. 124) that

> British politics are almost wholly innocent of those issues which cross the social lines in other lands – for example race, nationality, religion, town and country interests, regional interest, or the conflict between authoritarian and parliamentary methods.

and concluded his own analyses, with the statement that

> we have verified John Bonham's hypothesis . . . that very little except class matters for politics in Great Britain. In addition, we have discovered little evidence of a decline of the association of class and party (Alford, 1963, p. 170).

Alford's conclusions were based on analyses of opinion poll and survey data, to which he applied a relatively simple test. He assumed that parties of 'the Left' would have most support among the working class, whereas those of 'the Right' would draw most support from the middle class. This led to the derivation of an *index of class voting* defined as

> Subtract the percentage of persons in non-manual occupations voting for Left parties from the percentage of persons in manual occupations voting for Left parties (Alford, 1963, pp. 80-1).

The greater the index, the greater the inter-class polarisation. Thus if the results of a survey showed

	Per Cent Voting	
Class	Left	Right
Middle	10	90
Working	70	30

the index value would be 60 (70-10). Alford's reworking of data from eight national surveys (and votes over the period 1957-62), produced an average index of 40 (with the range between 35 and 44): the averages for the other three countries were – Australia, 33; United States, 16; and Canada, 8. Most of the surveys showed Labour obtaining around 65 per cent of the votes of the working class (the range was 57 to 68) and 25 per cent of the middle class votes (with a range from 19 to 37).

Although Alford's main finding is the one most often cited, his analyses produced several other important conclusions. First, he noted that Labour invariably does less well in winning middle-class votes than does Conservative in winning working-class votes. (Working-class conservatism has been of particular interest to researchers: see McKenzie and Silver, 1968; Runciman, 1966.) Secondly, he observed that the socially upwardly mobile have not retained their original Labour predispositions, but have shifted to become Conservative voters; thus

> the shift of the occupational structure has undermined the objective basis of Labour support by reducing the actual size of the working class (Alford, 1963, p. 133).

Thirdly, he concluded that rural location does not influence class voting trends, although in 'heavily Labour urban constituencies' (p. 133) the index of class voting is above average. Fourthly, religion although of some influence (Catholic working-class members are more likely to vote Labour than Protestants; the same is not true in the middle class), is of relative insignificance. Regional location, too, – excluding Scotland and Wales – does not have a very strong influence on voting:

> a region with economic problems predisposes its working class residents to vote especially strongly Labour, but whatever regional subculture may exist does not produce disproportionate Labour voting among the middle class (Alford, 1963, p. 151).

Finally, Alford refutes suggestions of a substantial age cleavage.

The British Election Study: Butler and Stokes

Alford's book was based on a reanalysis of opinion poll data, complemented by reference to local surveys by political scientists and sociologists. It was not until the 1960s that large-scale, national academic surveys were undertaken in Britain, comparable to those already frequently conducted in the United States. The original classic study was the sequence of surveys in 1963, 1964, 1966, and the follow-up sequence in 1969 and 1970, conducted by David Butler and Donald Stokes and reported in the two editions (1969, 1974) of *Political Change in Britain*.

These surveys, most of them conducted immediately after a general election, used sophisticated questionnaires designed to tap a wide range of attitudes and the careful sample design was used so that conclusions about trends could be drawn.

In the second (1974) edition of their book, Butler and Stokes focused in their conclusions on trends in electoral support during the 1960s. With respect to the long term, they focused on the dominant class-cleavage and the influence of an ageing and changing electorate; for the short term, they concentrated on the relevance of issues plus the positions of the parties and their leaders on these. One of their major conclusions was of increasing volatility in electoral behaviour, a theme developed further by the students of elections in the 1970s (see below).

Unlike Alford, Butler and Stokes did not identify continuity in the class-cleavage; their data provide an index of 47 for 1963 but 34 for 1970. Thus they concluded that

> the electorate has become progressively less inclined to respond to politics in terms of class and that the class appeals of the parties themselves have become much more muted (Butler and Stokes, 1974, p. 192)

They relate these changes to a variety of factors, including general social and economic improvement for the population and the record of the Labour Party when in power. A growth in 'working-class conservatism' was being countered by a rise in middle-class Labour support.

The trends noted above notwithstanding, Butler and Stokes' data and discussions illustrate the continuing strength of the class-cleavage. Other influences are present, and are illustrated, but the picture of 'one nation' in terms of electoral behaviour is sharp and clear. Regarding regional variations, for example, they note that the link between class and party exists in all English regions, although the levels of partisan support vary: in 1963-6, the index of class voting for northern and southern regions differed only slightly (41 and 38, respectively), but in the former Labour was supported by 71 per cent of the working class, compared to 63 per cent in the latter. They also showed – in ways that will be discussed more fully below (p. 64) – that

> once a partisan tendency becomes dominant in a local area

processes of opinion formation will draw additional support to the party that is dominant (Butler and Stokes, 1974, p. 129)

Geographical variations are accentuated.

Despite the findings reported in the previous paragraph, it is implicit throughout Butler and Stokes' books that Britain, and in particular England, is 'one nation' in terms of electoral behaviour. Class dominates all else. This inference comes through clearly in two other aspects of their treatment.

As a commentator on electoral trends in Britain, David Butler has been responsible for the widespread use of the concept of *swing* to describe changes between elections. Swing is defined as the difference between the percentage of the votes won by a party at two elections (i.e. the percentage points change). If there are only two parties, then the swings cancel each other. Thus, if the results in constituency i at two elections are

	Per Cent Voting	
Election	Conservative	Labour
1	60	40
2	63	37

then there was a three per cent swing to Conservative (63-60) and a three per cent swing away from Labour (37-40). With three parties, swing is less useful as an index of change, but it remains in popular usage usually referring to the votes won by Conservative and Labour only.

David Butler's analyses of swings at the constituency level led to his conclusion of an 'impressive evenness of swing' (Butler and Stokes, 1974, p. 121). The implication is that the volume of change was the same everywhere, giving further credence to the 'one nation' view of British politics. If the national trend to Conservative were three per cent (i.e. an increase of three percentage points), then the implication was that it was three per cent everywhere.

This conclusion would appear to be an overstatement, and the implication unwarranted. The data on which it was based were:

Inter-election Period	Mean Swing	Standard Deviation
1950-1	0.7 (to C)	1.4
1951-5	1.8 (to C)	1.4

1955-9	1.1 (to C)	2.2
1959-64	3.5 (to L)	2.4
1964-6	3.2 (to L)	1.7
1966-70	4.4 (to C)	2.1
1970-4(Feb.)	0.9 (to L)	2.9

(the data are in Butler and Stokes, 1974, p. 121). The conclusion of 'national uniformity' does not seem to be warranted, with a standard deviation at least twice the mean in three of the seven inter-election periods. Thus in the last period, for example (assuming a normal distribution), a swing of 6.7 percentage points or more to the Labour Party would have occurred in three per cent of the constituencies, and a swing of 4.9 percentage points to the Conservatives in a further three per cent.

In promoting the concept of uniform swing over the country, therefore, David Butler has disregarded a substantial amount of variation. (Butler and Stokes note 1974, p. 121 – that 'Of course, the generally uniform pattern of swings between the parties has always been broken by regional and local variations' but the focus on uniformity indicates the major thrust of their thinking, and the conclusion that others have drawn from their work.) There is substantial evidence of continuity in the geography of voting in England over a long period (Johnston, 1983a), but this is not equivalent to a uniformity of swing. Further, as Butler and Stokes (1974, pp. 140-51) demonstrate, percentage points change is not the same as per cent change, so that, even if there were uniformity of swing this would not imply similar shifts in all places. (Instead, it would imply – necessarily according to Johnston and Hay, 1982 – that for a party gaining votes nationally the swing would be greater than average in areas where it was already strong and less than average where it was initially weak: Johnston, 1981a, 1981b). A lack of geographical variation is the consequence of geographical variation! (This apparent paradox is analysed more fully in Chapter 4, which shows that Butler's measure provides a poor index of change.)

The concept of swing represents the *net* changes in the pattern of voting between elections. Butler and Stokes also introduced a 'flow-of-the-vote' matrix, which shows the *gross* changes. (The net changes can be derived from aggregate data – i.e. election returns; the gross changes can only be derived from survey data.) Such matrices show the number of voters who remained loyal to party between two elections, the number who shifted their allegiance, and so on. Table 1.1 shows the matrix for the inter-election period 1966-70. Thus, of those who voted Conservative in 1966, 71.3 per cent (19.1/26.8) remained loyal in 1970, whereas 3.0 per cent shifted allegiance to the Labour Party. The latter retained the support of only 57.3 per cent of those voting for it in 1966; 10.2 per cent shifted allegiance to the Conservative Party whereas 20.7 per cent abstained in the second election.

Table 1.1 The 'Flow-of-the-Vote', 1966-70

1966 Vote	1970 Vote[a] Conservative	Labour	Liberal	Abstain	Left System[b]
Conservative	19.1	0.8	0.4	3.7	2.8
Labour	3.2	18.0	0.8	6.5	2.9
Liberal	1.5	0.7	2.0	1.1	0.3
Abstain	3.5	3.5	1.1	12.1	1.0
Outside System[b]	3.5	5.9	0.8	4.8	

Source: Butler and Stokes (1974, p. 263).
Notes:
a. Expressed as a percentage of the table total.
b. Those outside the system in 1966 were those not qualified to vote (mainly by age) then; those who had left the system by 1970 had either died or emigrated.

Construction (using remembered voting) of matrices for previous periods allowed Butler and Stokes to identify increased volatility in the electorate between 1959 and 1970. Whereas 59 per cent remained loyal in the early years, only 44 per cent did later in the decade. The amount of turnover involved a minority of electors, however. If the matrices for each period were summed, then only 49 per cent of those voting Conservative in 1959 would have voted the same way in 1970. The actual figure was 68 per cent, suggesting that the electorate is divided into a loyal majority and an (increasingly substantial) volatile minority, with the latter changing their preferences frequently.

The analysis of the 'flow-of-the-vote' matrix makes no reference to possible variations in its composition – from social group to social group or area to area. (Johnston and Hay, 1982, have shown that, given the geography of voting in Britain, such variations must have been present: see also Johnston, 1981a.) The clear implication is that the gross pattern shown by the matrix is not only the national situation but is typical (within the bounds of sampling error) of all parts of the country. Butler and Stokes' analyses of the causes of volatility make no reference to *where* the volatile voters live.

The British Election Study: Sarlvik and Crewe

For the three elections of February 1974, October 1974, and May 1979 the British Election Study was transferred to the University of Essex. Very similar questionnaires and analyses were used, and the same 'one nation' inference can be drawn from the writings of the principal investigators.

By the end of the 1970s, the growth in support for the Liberal Party demanded that it be included in the tabulated data. In general terms, it drew support about equally from the two classes, as 1979 data show (Sarlvik and Crewe, 1983, p. 81):

| | Per Cent Voting | | |
	Conservative	Labour	Liberal/Other
Middle class	60	23	17
Working class	35	50	15

These give an index of class voting of only 27, representing a decline since 1974 as a result of increased support for the Conservative Party among the working class. (This is what Crewe, 1981, p. 279, identified as Labour losing the election because of a 'massive hæmorrhage of working-class votes'; elsewhere – Crewe, 1982 – he was suggested that this has been brought about by conflicts within the working class, such that the Labour Party is no longer able to represent the interests of that class as a whole in its policies.)

Given the decline in the importance of the class-cleavage, Sarlvik and Crewe investigated other sources of variation in support for the parties. Their analyses paid scant regard to geographical variations – only one variable was included; a north: south split of the country (with Greater London being included in the north!). For both Conservative and Labour, they found that occupational class was the best predictor, but this accounted for only 9 and 10 per cent of the variation respectively. Other important predictors (accounting for at least 5 per cent of the variation) were: membership of a trade union; housing tenure; and living standard. In general, the view of a national (through increasingly weaker) class-cleavage is upheld: however

> the voters' opinions on policies and on the parties' perform-
> ances in office 'explain' more than twice as much as all the
> social and economic characteristics taken together (Sarlvik and
> Crewe, 1983, p. 113)

The link between party and class is now mediated through
perceptions of party performance which are only imperfectly
related to class.

Just as where the voters live is almost entirely absent from
Sarlvik and Crewe's analyses of the voting cleavages, so it
receives no attention in their consideration of the 'flow-of-the-
vote'. Their detailed calculations lead them to identify

> a marked degree of flux in the British electorate. When a third
> of the electorate switches votes between two elections, and up
> to half change vote at least once in the decade, we can no
> longer regard the 'floating voter' as confined to a small
> minority (Sarlvik and Crewe, 1983, p. 73)

And yet Sarlvik and Crewe show even less interest in who the
floating voters are, and where they live, than did Butler and
Stokes. The greater the volatility, the stronger the implication
that it is the same everywhere.

And 1983

The 1983 British Election Study was conducted at the University
of Oxford. At the time of writing (late 1983) no data or analyses
had been published. At the time of the election, however, the
BBC commissioned a Gallup Poll of some 4,141 voters; the
results were analysed within a week of the election and published
in *The Guardian* (Crewe, 1983a, 1983b).

The tenor of these analyses mirrors that of all the previous
output of the British Election Study. The class-cleavage is given
pride of place (despite the finding of Sarlvik and Crewe that it
accounts for only 10 per cent of the variation), and there is no
spatial disaggregation of either the class-cleavage or the 'flow-of-
the-vote' matrix. Thus, despite the Alliance performing better
among women than men voters, and best of the three parties with
the new voters

> It is not age or sex, but social class, however defined, that continues to structure party choice (Crewe, 1983a, p. 5)

Labour's support among the working class was further eroded, especially among the owner-occupiers and non-members of trade unions:

> the Labour vote remains largely working class; but the working class has ceased to be largely Labour

Alternatives to Class

Two conclusions stand out from the work reviewed above – most of it related to the British Election Study. The first is the primacy given to the class-cleavage. The second is the general implication that place is unimportant: where voters live does not influence how they vote.

Regarding the first of these conclusions, it could be argued very strongly that the evidence, including that used by the authors cited, does not justify the continued importance attached to class. Pulzer (1967, p. 98) wrote that

> Class is the basis of British party politics: all else is embellishment and detail.

Sixteen years later, analysts make similar statements; statements which appear increasingly invalid.

The case against the 'class-equals-party' model has been developed by Richard Rose (1974), and recently emphatically restated (Rose, 1982a). He shows that the relationship between class and party has declined substantially in recent decades, in part because the simple, ideal type of model of classes in Britain is no longer relevant.

> In Britain, an ideal-type manual worker, in addition to a manual occupation, is expected to have left school at the minimum legal leaving age, to belong to a trade union, live in a state-owned council house, and subjectively identify as working class. Reciprocally, a middle-class person, in addition to having a nonmanual occupation, is expected to have more

than a minimum of education, to be a homeowner, not belong to a trade union, and subjectively identify with the middle class (Rose, 1982a, p. 150)

Few fit these ideal types. According to Rose, in 1964 only 14 per cent of the electorate were 'ideal typical' working class, and in 1979 the figure was only 4 per cent; for the middle class ideal type, the percentage fell from 12 to 10. Thus almost 85 per cent of the electorate is not readily classifiable as working or middle class. This suggests a much more complex pattern of cleavages, and the importance of foundations other than class position in the formation of political attitudes. (On the latter, see Himmelweit *et al.*, 1981; see also Dunleavy's, 1982, critique.)

Rose analysed the influence of ten separate variables on party choice: occupation, housing tenure, union membership, age, religion, sex, education, nation of residence, car ownership and telephone ownership. For data from the 1959 British Election Study he found that these ten together accounted for 21 per cent of the variation in party choice. By 1979, they accounted for only 12.1 per cent. The relevance of the class variable (occupation) became almost nil: it accounted for 15 per cent of the variation in 1959 but only 0.7 per cent in 1979. Thus whereas, he says, 'journalistic sources would suggest that since 1959 class differences in British party politics have been increasing' (Rose, 1982a, p. 152), in fact class polarisation has substantially declined.

Rose suggests that the search for influences in who votes for which party must focus on the process of political socialisation. He suggests a sequence of eight stages to this, as follows:

(1) age;
(2) sex;
(3) father's class;
(4) father's party;
(5) religion;
(6) education;
(7) current class; and
(8) other contemporary influences (housing tenure; nation of residence; union membership).

His analyses of 1964 and 1974 British Election Study data show that in both, the largest influences were those represented by stages 3 and 4, accounting for 22.6 per cent of the variation in 1964 and 19.1 in 1974. Current class came third in 1964 (7.7 per cent) but declined to fourth place (2.4 per cent) behind housing tenure (5.5 per cent) in 1974. Finally, Rose argues (as did both Butler and Stokes and Sarlvik and Crewe) that to predict voting at any one time macroeconomic conditions must be taken into account.

Rose's political socialisation model is not only more successful than the traditional class-cleavage model, it also bears more relation to the more open society that characterises contemporary Britain. But, apart from the reference to nation of residence (England vs. the rest: see Rose, 1982b; Madgwick and Rose, 1982), he ignores what to many people is a significant influence on socialisation – the environment within which it occurs. A similar criticism can be directed at the work of Franklin (1982). In this, six variables were used to characterise the voters in the survey: three representing childhood political socialisation (parents' class; parents' party; education); and three representing the voter's current situation (occupation; union membership; housing tenure). The socialisation variables omit any reference to the wider environment outside the home (see Wright, 1977, for an example of their relevance), as do the three for current situation. Does the situation of other voters in the area influence an individual's political attitude-formation?

A final set of, trenchant, criticisms of the 'class-equals-party' model has been provided by Dunleavy (1979, 1980a, 1980b), who notes that

> The declining association between occupational class and political alignment in Britain has now been documented by a number of studies (Dunleavy, 1979, p. 410)

According to Crewe and others, he says, this declining association is the concomitant of partisan dealignment, a necessary conclusion because there has been 'no shift in the social structure . . . [producing] an enduring, nationwide realignment of party support since 1945' (Crewe, 1976, p. 46). On the contrary, Dunleavy argues that there have indeed been major changes in the social structure, associated with access to consumption goods and with the growth of public sector employment, and he contends that these 'extensive changes in

British society . . . could well have political effects cross-cutting those of occupational class' (Dunleavy, 1979, p. 410).

Dunleavy developed his critique by outlining a theory of the division of society into 'consumption sectors', arguing that access to certain consumption goods is now organised collectively rather than with reference to household income. A new cleavage has developed in British society, separating, on the one hand, those whose consumption is the result of individual decisions, made in the private market place, from, on the other hand, those whose consumption is organised collectively in the public sector. These cleavages form the basis of political choices, with the former group represented by the ideology of the Conservative Party and the latter by that of the Labour Party. Dunleavy tested the validity of this contention by reanalysing survey data to show that housing tenure and car ownership are related to the likelihood of voting Conservative, independent of occupational class. Furthermore, household type is also related to Conservative voting; members of family households (those with at least one child under 16, and therefore major beneficiaries from collective consumption) are less likely to vote Conservative than are members of all-adult and elderly households. Nevertheless, despite these important influences on voting which cut across the traditional occupational class-cleavage, Dunleavy's analyses show that the latter is still the main single determinant of party choice.

Dunleavy's analyses of consumption-related cleavages, and also of cleavages based on union membership (Dunleavy, 1980b), are entirely aspatial. This is not only an implicit rejection of any hypothesised relationship between residential location and partisan choice. Dunleavy (1979) is scathing in his rejection of such a hypothesis; to him, the where of political socialisation is apparently irrelevant. (Dunleavy's criticism is considered in more detail in Chapter 5.)

The Aggregate View

All of the studies cited above have been based on survey data from which, because of the relatively small numbers of observations, locational influences are difficult to decipher (see Taylor and Johnston, 1979, ch. 5) – even if the investigators are interested in such influences. With aggregate data, on the other hand, locational influences are more readily identified, although the search involves the problem of the ecological fallacy (Taylor and Johnston, 1979, ch. 1), of reading into the data inferences about individual behaviour that may be unwarranted.

As noted above, aggregate (or ecological) data analyses of voting in Britain were not possible for general election studies prior to the release of 1966 Census data at the constituency scale. Using these data, it was possible for the first time to regress, for example, the Labour percentage of the vote by constituency against characteristics of population and households, and deduce conclusions about who voted for what. (Thus, for example, if the percentage voting Labour is positively related to the percentage of households in poor quality housing, it could be inferred that residents of such housing are more likely to vote Labour: the problems of such ecological inference are stressed by Crewe and Payne, 1976.)

The first sophisticated use of aggregate data analysis to investigate individual voting behaviour in Britain was conducted by Crewe and Payne. In an initial analysis (Crewe and Payne, 1971) of the 1970 general election they found a general relationship between the Labour vote (as a percentage of the votes for Labour and Conservative combined) and the percentage of the male workforce (including the retired) in manual occupations (i.e. the working class). The residuals from this relationship suggested the influence of other variables, and later analyses (Crewe and Payne, 1976) used a more sophisticated model that allowed them not only to predict the Labour vote very successfully but also to estimate the percentage of manual workers who voted Labour in various types of constituency. They found not only that the Labour vote increased with the relative size of the working class in a constituency, but also that: it was less in seats with relatively large agricultural workforces; it was greater in seats with substantial numbers of miners; it was greater in safe Labour seats and less in safe Conservative seats; and it was greater where other parties did well at the apparent expense of the Conservative Party. Thus, as a result, the percentage of manual workers estimated as Labour voters in 'very Labour' seats was about 81, whereas in 'fairly Conservative' seats it was only 44.

The results of such a careful analysis clearly indicate the need to incorporate locational influences in any model of who votes for which party. Thus Crewe and Payne (1976, p. 67) argue that there is

> the possibility that when Labour voters are in a small minority, they are *less* loyal than Conservatives in the same situation

and also that (p. 79)

> strong Labour areas may be politically and socially uncongenial to workers voting Conservative.

They also conclude that it is possible to predict the Labour vote very closely using only a few variables. Of these

> Social class in the form of 'percent manual workers' is easily the strongest single predictor (p. 67).

Used simplistically, this conclusion can be interpreted as substantial support for the 'class-equals-party' model.

In other work, Crewe (1973) extended the treatment by inquiring into differences within the working class in the level of Labour support. According to the embourgeoisement thesis (Goldthorpe *et al.*, 1968), affluent members of the working class should be more prepared to vote Conservative – reflecting their position (e.g. as home owners) within society. Crewe found some support for this thesis (as also did Garrahan, 1977, with regard to occupational class and housing tenure). Later work by Johnston has shown that the Conservative vote in 1971 in English constituencies was positively related, *ceteris paribus*, to car ownership rates within the working class (Johnston, 1981c), and that there was a significant difference within the working class between car-owners and non-car-owners in their propensity to vote Conservative, but not within the middle class (Johnston, 1982d).

The aggregate approach has also been used extensively by Miller, in a number of studies. One of his main goals was to create a data set whereby electoral and census data could be jointly analysed for a long time period, and he produced 161 'constant units' (Miller, Raab and Britto, 1974) that could be used for every election in the period 1918 to 1974. His analyses of England suggested voting patterns very similar to those identified by Crewe and Payne (1976); the dominant influences on a party's votes were the class composition of an area, the relative importance of agricultural occupations and the religious composition of the population.

In addition to his analyses of the 1918 to 1974 period, Miller also conducted a more detailed analysis of the 1966 general election. Again, and not surprisingly, his results were very similar to those of Crewe and Payne. Miller, too, noted that there were substantial inter-constituency variations in the percentage of the members of a given class voting for a particular party. For example, regressing Conservative vote on middle-class (i.e. non-manual) occupations he obtained:

from survey data
$$C = 0.300 + 0.434 \, M$$
from constituency data
$$PC = 0.160 + 0.930 \, PM$$

where

    C = 1 if an individual voted Conservative, 0 otherwise;
    M = 1 if an individual was in the middle class, 0 otherwise;
    PC is the proportion who voted Conservative in a constituency; and
    PM is the proportion of a constituency's workforce who are in the middle class.

In other words, whereas a member of the middle class was likely to vote Conservative at a probability of 0.764 according to the survey data, and a member of the working class had a probability of 0.3 of voting Conservative, according to the ecological analysis in a constituency comprising all middle-class voters, all of them would have voted Conservative, whereas in one comprising all working-class voters, only 0.16 would have voted Conservative. As Miller (1977, p. 65) argues, this suggests that

> The partisanship of individuals is influenced more by where they live than what they do.

Middle-class electors are more likely to vote Conservative than Labour; middle-class electors living in middle-class areas are even more likely to vote Conservative.

In later work, Miller has extended his analyses of these locational effects, tackling the issue of dealignment identified by the survey analysts (Butler, Crewe, Rose, Stokes, etc.) discussed above. Miller (1978) suggests that society should be divided into 'core' classes – similar to Rose's (1982a) ideal types – clearly identified with the two main parties: for Conservative, the core comprises 'employers and managers', whereas for Labour it is manual trade unionists. The members of these classes, he argued, were 'politically insensitive to their environments' (p. 277), and members of such cores tended to vote the same way, wherever they lived. Others tended to vote according to the relative importance of these core classes in their constituency.

The major determinant of constituency class polarization is . . . the environmental effect of the local balance of core classes (p. 281)

Miller (1979) had shown that this remained the case at the 1979 general election. Survey data has shown a declining difference between the classes in their partisan support; analysis of constituency data showed that interclass polarisation remained, based on the role of the dominant core class in each area. (Curtice and Steed, 1982, have presented a similar analysis, based on urban:rural and north:south differences.)

Work using aggregate data has paid more attention to geographical variations in voting, therefore. This has been continued in the initial analyses of the 1983 general election. Berrington (1983), for example, has shown that the Conservative Party gained most ground over the Labour Party, relative to their 1979 performances, in safe Conservative seats, and least in safe Labour seats (suggesting the increase in spatial polarisation identified by Curtice and Steed, 1982). Similarly he notes regional and urban:rural variations in which

the general tendency has been to confirm the geographical division of the country which has been developing since 1955 in which the north and the conurbations have become more Labour, and the more prosperous south, the suburban and semi-rural areas, and the countryside, more Conservative (p. 267)

Steed and Curtice (1983) have focused on the performance of the Alliance, though their regression analyses confirm that the usual correlates of Conservative and Labour voting were again present at the constituency level. They conclude that

> the Alliance is stronger in middle-class areas, in agricultural areas, in resorts . . . and in expanding areas . . . two areas of characteristic Alliance weakness [are] measured by higher unemployment and higher electoral density (p. 9)

More detailed analysis suggests that

> the Alliance's gains should prove greatest amongst those who are neither particularly middle class nor the working class suffering from or fearing depression (p. 10)

and the distinct section of the middle class who are pro-Alliance comprises, they believe, 'those who owe their social position to economic achievement rather than to qualifications, or what we might call the business middle class' (p. 12). In addition, they identified a number of regional and local factors.

In Summary

Two basic beliefs underlie much popular and academic presentation of electoral cleavages in England: that class is the dominant influence on party choice; and that class is declining in importance with a breakdown of the polarisation within society. (The Alliance seeks to exploit the latter, arguing that Conservative and Labour reflect an obsolete, polarisation view of society.) Some argue that class is being replaced by other cleavages (Dunleavy); some argue that political socialisation is more complex than can be represented by a single variable (Rose; Franklin); and some suggest a decline in the predictability of electoral choice (Crewe).

These analyses have one characteristic in common: they pay little attention to the importance of place as an influence on voting, whether in the longer-term processes of political socialisation or in providing the immediate context at the time of a particular election. It may be no accident that they also rely almost entirely on survey data, from which locational influences are difficult to determine.

Countering this 'class-equals-party'/dealignment position are the results of two sophisticated analyses of aggregate data which suggest very clearly that where people live is an important influence on how they vote. Such findings tend to be ignored, however. The general picture of voting in England suggests that its geography is unimportant. The remainder of this book seeks to bolster the argument that it is not.

2 GEOGRAPHICAL VARIATIONS IN VOTING

The general thesis of this book is that location has been substantially ignored as a variable influencing voting behaviour in England. 'Who votes what?' is widely considered a sufficiently complex question for electoral analysis; the contention here is that 'who votes what, where?' is the question that should be asked.

To promote this point of view, the present chapter presents two sets of analyses. The first looks at the results of the 1983 general election, and argues that the pattern is strongly suggestive of a major locational component to party choice. The second enquires whether this pattern can be accounted for as merely a mapping of the answer to the question 'who votes what' at the national scale.

The 1983 Result in England

The result of the 1983 general election, in terms of the composition of Parliament and the creation of a government, was an overwhelming victory for the Conservative Party. With some 43 per cent of the national (i.e. British) vote it obtained nearly 63 per cent of the seats. For the other two major parties/groups, defeat in the popular vote was magnified by the allocation of seats. The Labour Party won 28 per cent of the votes, and yet won 33 per cent of the seats, but the Liberal/SDP Alliance was only two percentage points behind Labour in the contest for votes yet obtained only 3.6 per cent of the seats.

This mismatch between votes and seats is a characteristic feature of British election results. It is a consequence of the interactions between the nature of the electoral system and the geography of party support, with the procedures used for defining the constituencies adding to the cause of the mismatch.

The reason for this electoral bias has been analysed in detail elsewhere (the most detailed and sophisticated work being that of

Gudgin and Taylor, 1979). The British electoral system involves a division of the country into a set of territorial constituencies, each of which returns one member to the House of Commons. The victor in each constituency is the candidate with the largest number of votes, irrespective of whether they form a majority of the votes cast let alone of those that could be cast. The larger the number of candidates, the smaller the percentage of the votes that could produce success. (The minimum percentage, in integer terms, is ($[100/n] + 1$), where n is the number of candidates. With four candidates, 26 per cent would be enough for victory if the distribution were 26, 25, 25, 24, and with six 17 per cent would be sufficient in certain circumstances: see Loosemore and Hanby, 1971.)

Given the 43:28:26 distribution of votes among the three main parties/groups in 1983, therefore, it would have been possible for the Conservative Party to win every seat, if that distribution had been repeated in each constituency. The greater the deviation from the national figure in individual constituencies, the greater the likelihood of a different result in terms of seats. As Gudgin and Taylor (1979, see also Taylor and Gudgin, 1982) have shown, the greater the variance in a party's percentage of the vote across a set of constituencies, the greater its potential for winning at least some seats, given that it wins less than half of the votes overall. Thus a party with a low percentage of the votes and a low variance may win few seats: a party with a low percentage but a high variance may win some seats; and a party with a high percentage of the votes and a low variance may win a large proportion of the seats.

The validity of this conclusion is shown by Figure 2.1, which contains the distributions of vote percentages, by English constituency, in 1983 for each of the three parties. The means and standard deviations related to these are:

	Mean	Standard Deviation
Conservative	45.73	11.68
Labour	28.04	15.06
Alliance	26.24	6.95

In relative terms, therefore, the Conservative Party had a high mean and a low variance, the Alliance had a low mean and a low variance, and Labour had a low mean and a high variance. The

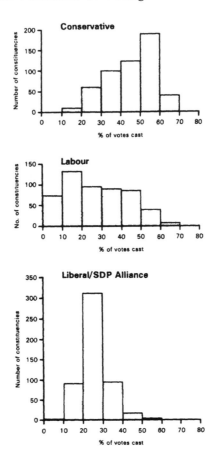

Figure 2.1 Frequency Distributions of the Percentage of the Votes Cast for Each Party, England, 1983

Conservative landslide in terms of the allocation of seats was a clear consequence of these three distributions.

To win a seat, a minimum of 34 per cent of the votes is needed (assuming only Conservative, Labour and Alliance candidates). For the Alliance, this was a rare event; in the majority of constituencies, the Liberal/SDP candidates polled between 21 and 30 per cent of the votes, sufficient to come second in many but insufficient to bring victory. For Labour, on the other hand, there was a substantial number of constituencies in which the party won more than the minimum threshold of votes necessary for victory; whether it won a seat with 34 per cent or more of the votes depended, in those where it won less than half, on the distribution of the remainder between Conservative and Alliance. In some it did not, but in most of those where it had at least 40 per cent of the vote it won. Compared to the Alliance, Labour was strong enough in certain areas to win seats. And so too, of course, was the Conservative Party, in many parts of the country.

The basic reason for the differential conversion of votes into seats is thus the geography of support for the contestants. Figure 2.1 clearly shows that the Alliance won strong minority support nearly everywhere, but substantial support in very few places; approximately one-quarter of the voting electorate in all parts of the country preferred the Alliance candidate. Labour, too, won only minority support, but the geography of that support was much more variable. It, too, obtained the votes of little more than one-quarter of the electorate, but in 25 per cent of the constituencies it obtained 41 per cent of more of the votes (compared to only 4 per cent of the constituencies for the Alliance); countering this, Labour won 20 per cent or less in 39.4 per cent of the seats. Finally, for the Conservative Party, the slightly negatively-skewed distribution in Figure 2.1 shows that it rarely won an overwhelming percentage of the votes; it was more likely to perform badly than very well, but was in general consistently winning about half of all the votes cast.

Two straightforward explanations can be offered for these geographies. The first is that they are faithful reflections of the social profiles of the electorate in the constituencies; the people who support Conservative are fairly evenly distributed across the country, as are those who support the Alliance, whereas Labour's voters are spatially more concentrated. The second explanation is that the parties have separate territorial heartlands, irrespective of the socio-economic composition of the electorate.

It is the first of these potential explanations that is implicitly favoured by most electoral analysts in England. Given the 'class-equals-party' assumption that underpins their analyses, then any major variation in the support for a party must reflect the geography of its class base. Thus it is assumed that the frequency distributions shown in Figure 2.1 reflect similar frequency distributions for the classes that support the different parties.

The validity of this assumption can be tested by constructing the relevant frequency distributions, using census data. According to the initial analyses of the 1983 election (Crewe 1983a, 1983b), the Conservative Party draws its main support from the white-collar occupational classes and from owner-occupiers, whereas Labour draws almost entirely from the manual (working) classes and the council tenants; the Alliance draws approximately equally from all categories. (The detailed data for England are given in Tables 2.1 and 2.2.) The distributions for these groups, taken from the 1981 Census data, are shown in Figures 2.2. and 2.3.

Table 2.1 Occupation, Social Class and Vote: England, 1983 (percentages)

Occupation	Conservative	Labour	Alliance	Other/Abstain
Professional-managerial	49.8	10.0	21.2	19.0
Other non-manual	42.9	17.4	17.0	23.7
Skilled manual	27.2	26.2	17.9	28.7
Other manual	18.9	31.6	16.8	32.7
Social Class				
I + II	51.7	10.2	20.2	17.9
IIIN	45.1	15.7	17.9	21.3
IIIM	27.6	24.2	18.2	30.0
IV + V	19.3	31.3	17.7	31.7
All non-manual	48.0	13.2	18.9	19.8
All manual	23.4	27.8	18.0	30.8

Source: Computed from the data tape of the BBC/Gallup election survey.

Table 2.2 Occupation, Housing Tenure, and Vote: England, 1983 (percentages)

Housing Tenure	Conservative	Labour	Alliance	Other/Abstain
Owned outright	47.9	12.7	18.7	20.7
Mortgaged	43.6	13.9	20.9	21.6
Council tenant	16.3	36.2	16.0	31.5
Other tenant	31.5	18.9	16.8	32.8
Other	26.3	18.4	26.3	29.0
Occupation/tenure				
Non-manual, own/mortgage	53.3	9.8	19.8	17.1
Non-manual, rent	32.7	23.0	16.5	27.8
Manual, own/mortgage	34.3	18.1	20.5	27.1
Manual, rent	16.4	34.1	16.4	33.2

Source: Computed from the data tape of the BBC/Gallup election night survey.

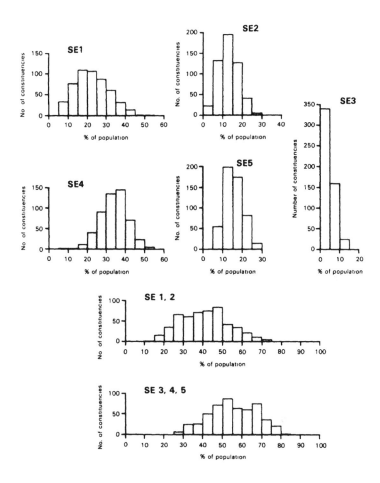

Figure 2.2 Frequency Distributions of the Percentage of the Population in Each Socio-economic Category, England, 1981

Tables 2.1 and 2.2 show that manual workers and tenants were most likely to vote Labour in 1983, wheras non-manual workers and owner-occupiers were most likely to vote Conservative. Given these sectoral cleavages, then Figures 2.2 and 2.3 indicate the variability in the inter-constituency distributions of the underlying bases of English voting then. Figure 2.2 shows that white-collar (or middle-class) voters form between one-third and one-half of the electorate in a majority of the constituencies, whereas manual (or working-class) voters mostly provide between one-half and three-quarters of the electorate. From the voting data in Table 2.1, it can be deduced therefore that the

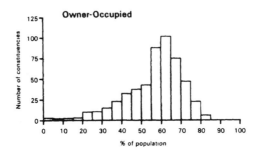

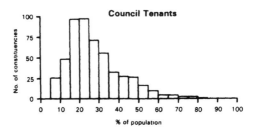

Figure 2.3 Frequency Distributions of the Percentage of the Population in the Two Major Tenure Categories, England, 1981

Labour Party won as many seats as it did not because it gained a majority of the working-class votes but because the working-class voters formed such a substantial majority in some constituencies that Labour was almost certain to win the seat. This is bolstered by the distribution of the two main tenure categories (Figure 2.3). Whereas residents of owner-occupied households are in the majority in most constituencies, council tenants form a large majority in some. Again, these concentrations of likely Labour voters probably account for that party's relative success (in terms of winning seats) at the 1983 election. Nevertheless, it is not clear that this was necessarily the case, and the next section investigates whether the geography of the election result was a faithful reflection of the geography of social and housing class.

Who Voted Where?

Surveys of the British electorate in 1983 suggested a continued difference between the classes in their level of support for the

Conservative and Labour Parties, therefore, and a loss of votes from both, especially Labour, to the Alliance. Did these trends repeat themselves in all parts of the country, and so produce the geography of the 1983 election result?

Class, Tenure and Vote

Social class is traditionally represented as the major determinant of voting behaviour in Great Britain – especially in England, where neither religion nor nationalist sentiment has any substantial impact. Class can be measured in a variety of ways, although the commonest using census information – given the absence of any questions relating to income – is occupation. Two scales are usually employed. The first comprises what the census authorities term *socio-economic groups*, of which there are seventeen at the present time (OPCS, 1980).

1. Employers and managers, large establishments;
2. Employers and managers, small establishments;
3. Professionals, self-employed;
4. Professionals, employees;
5. Intermediate non-manual workers;
6. Junior non-manual workers;
7. Personal service workers;
8. Foremen and supervisors – manual;
9. Skilled manual workers;
10. Semi-skilled manual workers;
11. Unskilled manual workers;
12. Non-professional self-employed;
13. Farmers – employers and managers;
14. Farmers – own account workers;
15. Agricultural workers;
16. Members of the Armed Forces; and
17. Others.

In this, people are allocated to a group according to their occupation and status. In the other, *social class*, people are allocated to one of six classes according to their occupational skills, viz:

I   Professional, etc.;
II   Intermediate;

IIIN Skilled non-manual;
IIIM Skilled manual;
IV Partly skilled; and
V Unskilled.

(The use of IIIN and IIIM indicates differences in types of skill, rather than levels, between non-manual and manual occupations.)

Both of these means of measuring social class were employed in the BBC/Gallup 1983 Election Survey (SSRC Data Archive, 1983). The occupational scale contained seven categories (with the OPCS equivalents in brackets):

1. Professional (3,4);
2. Director, proprietor, manager (1,2,13,14);
3. Shop, personal service (5,7);
4. Office, student, etc. (6);
5. Skilled manual (8,9);
6. Semi-skilled manual (10,15); and
7. Unskilled manual (11).

The social class scale contained six categories as in the census.

Respondents in the BBC/Gallup survey were asked their voting intention/choice, and this has been cross-tabulated against both occupation and social class in Table 2.1. (In Table 2.1, and throughout the book, the seven categories above have been collapsed to four (1+2; 3+4; 5; 6+7).) These tabulations show the typical pattern within the British electorate, with the non-manual classes strongly favouring the Conservative Party and the manual classes both substantially more pro-Labour and much more likely to abstain. The new element, the Alliance, picked up votes very evenly across all occupations and classes, with but a slight edge in the non-manual classes.

Housing tenure was placed into five categories in the 1983 survey. Cross-tabulated by vote (Table 2.2), the expected patterns emerge. Owners (including those purchasing their homes with a mortgage) are most likely to vote Conservative, whereas tenants, especially Council tenants, are more pro-Labour and more likely to abstain. The Alliance did slightly better among owners/buyers than among tenants.

Combining both occupation and tenure gives the fourfold

classification of respondents in the bottom panel of Table 2.2. (In this, outright owners and buyers were combined, because they are not differentiated in the census data. Tenants were combined with the small residual category to provide the other tenure.) With regard to the traditional main parties, this shows that residents of non-manual, owner-occupier households were most likely to vote Conservative (just over half of all the voters in that category) whereas residents of manual, tenant households were most likely to vote Labour (although only just over one-third of the members of that category voted Labour); the two polar groups also displayed major differences in the propensity to abstain. These figures confirm the traditional stereotype of English voting behaviour; between the two polar groups, those whose housing tenure was not consistent with the stereotyped occupation class fell between the two poles in their voting behaviour, with non-manual renters more likely to vote Labour than non-manual owner-occupiers, and manual owner-occupiers more likely to vote Conservative than their counterparts in tenanted property. The Alliance obtained votes in almost equal proportions in both occupational classes, but was three-four percentage points more successful among owner-occupiers.

Using these data on the basic electoral cleavages in English society as a whole, the election result was then 'predicted'. From the census tapes, the numbers of residents in each occupational category, each social class and each housing tenure category were obtained for every constituency, and these were summed to provide the figures for England as a whole. When the percentages in the various rows of Tables 2.1 and 2.2 were applied to the national row totals, and the resulting column totals derived, it was found – as in most other surveys – that the numbers of non-voters were understated (i.e. more people claimed that they had voted/intended to vote than turned out to be the case); as a consequence, voting for the parties was in some categories over-estimated. This meant that when the percentages in Table 2.1 and 2.2 were applied to the national totals in each voter category, the 'predicted' distribution of votes varied slightly from the actual. To remove this variation, the matrices were 'smoothed', using the entropy-maximising procedure outlined in Chapter 5, so that the average deviation between predicted and actual was below 5 per cent (see Appendix 3 for more details). The resulting matrices are given in Table 2.3.

Table 2.3 The 'Smoothed' Cross-classifications (percentages)

| | Vote 1983 | | | |
	Conservative	Labour	Alliance	Did Not Vote
Social Class				
I + II	51.3	9.1	20.6	19.0
IIIN	44.9	14.0	18.4	22.7
IIIM	27.6	21.7	18.7	32.0
IV + V	19.4	28.2	18.3	34.0
Occupation/tenure				
Non-manual, own/mortgage	49.9	10.7	20.7	18.8
Non-manual, rent	29.5	24.2	16.7	29.6
Manual, own/mortgage	31.1	19.1	20.8	29.0
Manual, rent	14.4	34.9	16.2	34.5

The two matrices in Table 2.3 were used to predict what the election result would have been in each constituency if the national pattern of voting had been repeated there. The

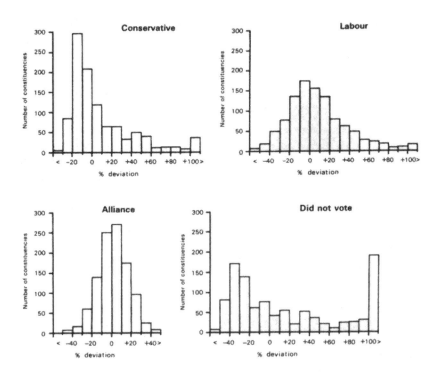

Figure 2.4 Frequency Distributions of the Percentage Deviations: Predicted Vote According to Social Class

differences between the actual and predicted vote for each party were then calculated, and expressed as percentages of the actual figure. If the national pattern of voting by social class and by occupation and tenure had been reproduced in every constituency, these deviations would have been very small. But they were not; the national pattern of voting did not reproduce the election result in the constituencies.

Figures 2.4 and 2.5 show the frequency distributions for these percentage deviation figures. In both, the striking feature is the range of deviation, especially with regard to voting for the Labour Party. Thus in Figure 2.4, in total only 31 per cent of the predictions using social class are within 10 per cent of the actual figure; for Labour the figure was only 11.3 per cent, and in 95 of the 523 constituencies the difference between the predicted and actual figure exceeded +100 per cent of the latter (i.e. the actual vote was less than half that predicted). The same pattern is repeated when occupation plus tenure is used to predict the vote

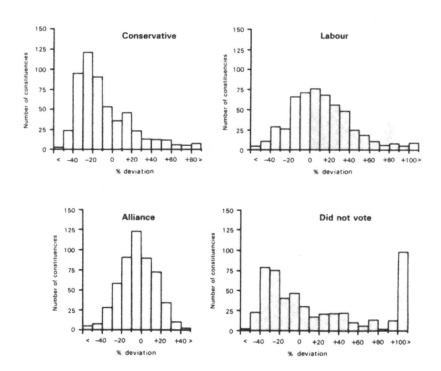

Figure 2.5 Frequency Distributions of the Percentage Deviations: Predicted Vote According to Occupation and Tenure

(Figure 2.5); again, the closest fits are for non-voters, with a deviation of less than 10 per cent of the actual in 41 per cent of the constituencies. (Note that the Conservative vote is predicted more accurately using occupation plus tenure than it is using social class alone.)

Where are the constituencies where the vote is not readily predicted from the national figures? Figures 2.6-2.9 show the upper and lower quartiles for four of the distributions. (All maps for constituencies use the same cartogram: see Appendix 1.) For

22 percent or more overpredicted

13 percent or more underpredicted

Figure 2.6 Percentage Deviations from the Actual Vote: Conservative Vote Predicted by Social Class

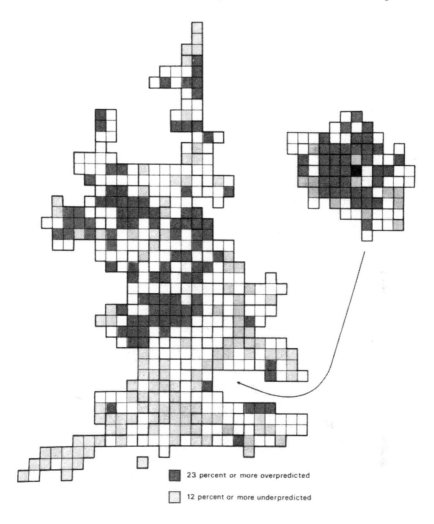

23 percent or more overpredicted

12 percent or more underpredicted

Figure 2.7 Percentage Deviations from the Actual Vote: Alliance Vote Predicted by Social Class

Conservative voting predicted from the social class matrix the pattern of deviations is clear (Figure 2.6); the main overpredictions are in the three northern regions, in Notts./Derbyshire, Stoke-on-Trent, in parts of the West Midlands and in Greater London, whereas the main body of underpredictions is along the south coast, in much of East Anglia, and in a few rural constituencies in the north. For the Alliance, again using social class as the predictor variable, the West Midlands, western

Figure 2.8 Percentage Deviations from the Actual Vote: Labour Vote Predicted by Occupation and Tenure

London, parts of Greater Manchester and Merseyside, Leicester, Nottingham and the South Yorkshire coalfield are the main areas of substantial overprediction (Figure 2.7), whereas there is a solid block of underpredictions in the Southwest.

The other two maps show the deviations from the predictions produced by the occupation plus tenure matrix. Figure 2.8 displays a clear north:south pattern for Labour, with under-

Figure 2.9 Percentage Deviations from the Actual Vote: Non-voting Predicted by Occupation and Tenure

predictions in the north and overpredictions in the south: in the north, only in a few rural constituencies plus resort-cum-retirement centres, such as Southport, is the Labour vote substantially overpredicted, whereas in the south only in a few Greater London constituencies, in Oxford, in Bristol and in Ipswich was its vote substantially underpredicted. Finally, the map for non-voting (Figure 2.9) has as its main feature substantial underprediction throughout most of Greater London and the south coast.

The geography of voting in 1983, therefore, suggests very major inter-constituency variations in who voted what; the distribution of votes as cast is not accurately predicted from knowledge of the national class and consumption cleavages.

The Flow-of-the-Vote

What of the short-term changes in voting between elections; are these the same in all parts of the country or do they differ substantially from constituency to constituency? To answer this, flow-of-the-vote matrices for the country as a whole are applied to the individual constituencies. (On the construction of these matrices, see Sarlvik and Crewe, 1983.)

Table 2.4 The Flow-of-the-Vote, 1979-83

A. From the Original Survey

1979	Conservative	Labour	1983 Alliance	Abstain
Conservative	77	4	13	6
Labour	7	63	22	8
Liberal	14	9	72	5
Abstain	22	12	14	52
Too young	28	17	20	35

B. Smoothed

	Conservative	Labour	Alliance	Did Not Vote
Conservative	73.9	3.5	11.2	11.4
Labour	7.1	57.9	21.0	14.0
Alliance	14.4	8.4	66.7	10.2
Did not vote	15.5	7.8	9.2	67.5

Source: Computed from the data tape of the BBC/Gallup election survey.

For the 1979-83 inter-election period, a national flow-of-the-vote matrix was derived from the same data file of the BBC Gallup Poll on election day, 1983 (initial results were published in *The Guardian*; Crewe, 1983a). This is given in Table 2.4A. It shows that party loyalty was greatest among 1979 Conservative voters (77 per cent of whom voted for that party in 1983) and among those who voted Liberal in 1979 (Alliance in 1983). There was a substantial flow of votes from Labour to the Alliance; like the Conservative Party, Labour lost three times more votes to the

Alliance than it did to the other 'major party'. The Alliance itself also suffered a considerable outflow of 1979 Liberal voters, with the Conservative Party benefiting most. Of those who either abstained in 1979 or who were not eligible to vote then, a large number (more than half of those in the former category) did not vote in 1983; of those who did, Conservative was the favoured destination. Crewe summarises these movements into four main trends: (1) the Alliance had a substantially greater impact on the fortunes of Labour than of Conservative; (2) there was a net flow of voters from Labour to Conservative; (3) more 1979 Labour than Conservative or Liberal supporters did not vote in 1983; and (4) Labour was disadvantaged by the votes of both those who abstained in 1979 and the first-time voters. In terms of the election result

> The main explanation for the Conservatives' win, therefore, is Labour desertions.

(Of course, as detailed in the next chapter, parties win elections in Britain according to the method of converting votes into seats.)

There is one column absent from Table 2.4A; no data are available on voters who left the electorate (mainly through death) between 1979 and 1983. Such information can only be obtained via panel surveys, which use the same sample after each election: the sample used for Table 2.4A was a cross-section. For the previous inter-election period (October 1974-9), Sarlvik and Crewe's (1983, p. 51) matrix suggests that 12, 10 and 7 per cent respectively of the Conservative, Labour and Liberal voters had left the electorate by 1979; this suggests that the Conservative voters were on average slightly older. (According to the 1983 survey – Crewe, 1983a – 48 per cent of those age 65+ voted Conservative, compared to 41 per cent of those aged 18-22.) Thus the matrix in Table 2.4A applies only to voters alive in 1983.

Applying that matrix at constituency level involves certain data problems. The first is the absence of information on either the number of deaths of electors in each constituency between 1979 and 1983 or the number of new voters who joined the electorate during the period. Although it would have been possible, using census data on the age distribution in each constituency in 1981, to derive estimates of these, it was decided that this time-consuming procedure was unnecessary. Consequently, it was assumed that deaths had no significant influence on the flow in particular places. Similarly, it was assumed that the patterns for new voters and abstainers were sufficiently similar, and the proportion of new voters in any one constituency insufficiently large to affect the flow substantially, so that they could be combined. Thus a new matrix was formed in which the first three columns and rows were as in Table 2.4A; the fourth column and row were entitled 'did not vote'. The sums of the first three columns and rows were thus those given in the election returns. The fourth column sum was the number of abstentions in 1983. The fourth row sum was the difference between the 1983 electorate and the total number of votes given to Conservative, Labour and Liberal in 1979; it combined the abstentions in 1979 with the new voters in 1983 (plus the small numbers of English electors who voted for other than the three main parties).

When the percentages in Table 2.4A were applied to the row totals, summing the columns produced a 56 per cent underestimate of the number of non-voters and 9, 15 and 16 per cent overestimates respectively of the Conservative, Labour and Alliance votes. Such a result is usual, as noted above. To correct for these deviations, the matrix was 'smoothed' (using the iterative procedure described in Appendix 3), so that no set of sums was on average more than 5 per cent different from the actual figure. The 'smoothed' matrix is given in Table 2.4B. The main difference is clearly in the flows into the 'did not vote' category. In relative terms, virtually no changes have been introduced, and Crewe's four conclusions remain valid.

To apply this matrix to each of the 523 constituencies in England, the row and column totals were needed. This introduced a further, major difficulty. Between 1979 and 1983 there was a redefinition of constituencies, and only 60 of the 523 were unchanged: changes of less than 5 per cent occurred in another 32. Since general election voting data are not reported at sub-constituency level, this means that the 1979 distribution of votes in a majority of the constituencies was unknown.

To circumvent these problems, a team of electoral analysts produced estimates of the 1979 voting in each of the 1983 constituencies (BBC/ITN, 1983). This was done using the results of local government elections in the constituent wards of the new constituencies. (The methodology is described by Fox in the BBC/ITN book.) These estimates were used to provide the party votes in each constituency in 1979. The 1983 results were known, and the total for the fourth ('did not vote') row in each constituency was derived as the 1983 electorate less the sum of the votes recorded in the first three rows. (Other researchers – see e.g. Steed and Curtice, 1983 – have used these estimates and found them valid for the vast majority of English constituencies; see Appendix 5.)

Using these 1979 estimated and 1983 actual vote totals for each constituency, the matrix in Table 2.4B was then applied, using the entropy-maximising procedure (Appendix 3), to the row totals to derive estimates of the 1983 vote there, assuming that the national flow-of-the-vote matrix applied. The estimate for each party and for the non-voters was then compared with the actual, and the difference was expressed as a percentage of the actual vote. Thus for Barrow and Furness the figures were

	Conservative	Labour	Alliance	Did Not Vote
A. Actual				
1983	22,284	17,707	11,079	16,826
B. Predicted				
1983	19,894	17,124	13,528	17,343
(B – A)	–2,390	–583	2,449	517
(B – A)/				
(A/100)	–10.7	–3.3	–22.1	3.1

The Conservative and Labour votes were underpredicted, the Alliance vote was substantially overpredicted, and non-voting was slightly overpredicted.

Over the full set of 523 constituencies, the figures of relative under- and overprediction indicate substantial differences: there can be little doubt that the national flow-of-the-vote matrix did not apply equally well in every constituency. The four frequency distributions are shown in Figure 2.10. For each of the three parties, the distribution is positively skewed, indicating a small number of very substantial overpredictions relative to the number of very substantial underpredictions. The range of values indicates many differences between predicted and actual that cannot readily be associated with the estimation procedure. This is particularly so with the Labour votes: in 25 per cent of the constituencies, the Labour vote was overpredicted by at least 50 per cent.

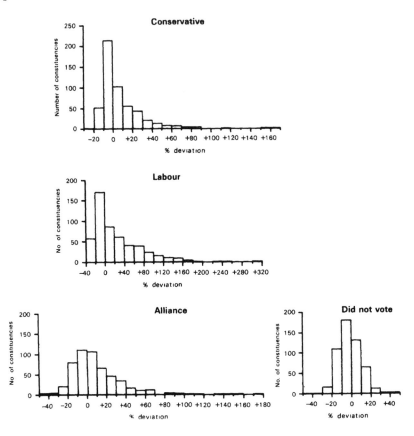

Figure 2.10 Frequency Distributions of the Percentage Deviations: Predicted Vote from 1979 Vote and the National Flow-of-the-Vote Matrix

The frequency distribution for non-voting is much closer to normal, and the range of deviations is much less than for the three parties. Overall, therefore, it must be concluded that whereas the distribution of electors between voters and non-voters in each constituency can be predicted quite readily from the national flow-of-the-vote matrix, the distribution of voters between the three parties cannot. This clearly suggests geographical variations in the switching of allegiances. What, then, is the geography of the deviations from the predicted values? Figures 2.11 to 2.14 show the constituencies in the upper and lower quartiles for each of the three parties and for the non-voters. (For a key to the cartograms used, see Appendix 1.) All four indicate clear spatial patterns.

Regarding the Conservative Party (Figure 2.11), there is a clear north:south differential, with Greater London providing the

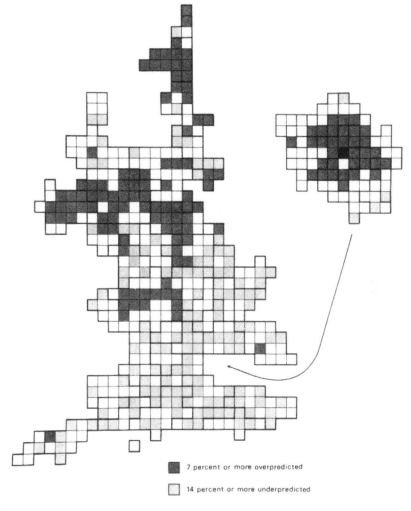

7 percent or more overpredicted

14 percent or more underpredicted

Figure 2.11 Percentage Deviations from the Actual Vote: Conservative Vote Predicted from 1979 Vote and the National Flow-of-the-Vote Matrix

only exception. Most of the constituencies in which the Conservative vote was 14 per cent or more overpredicted are in the north of England, and in the two southern conurbations of Greater London and the West Midlands. (The two exceptions are Thurrock, a Labour-won seat in Essex, and Plymouth, Devonport, the seat held by the present leader of the SDP.) Complementing this, most of the constituencies in which the

■ 13 percent or more overpredicted

□ 50 percent or more underpredicted

Figure 2.12 Percentage Deviations from the Actual Vote: Labour Vote Predicted from 1979 Vote and the National Flow-of-the-Vote Matrix

Conservative vote was underpredicted by 7 per cent or more are either in southern England or in the rural areas of the north.

Not surprisingly, the map for Labour (Figure 2.12) is to a considerable extent the mirror image of that for Conservative. The north:south differential is again very clear, with the Labour vote consistently underpredicted in the north and overpredicted in the south. In London, however, the Labour Party does not

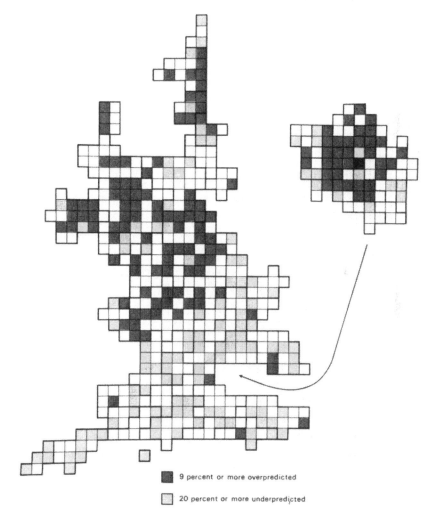

■ 9 percent or more overpredicted

□ 20 percent or more underpredicted

Figure 2.13 Percentage Deviations from the Actual Vote: Alliance Vote Predicted from 1979 Vote and the National Flow-of-the-Vote Matrix

seem to have benefited as much as might have been expected from the poor Conservative performance.

The Alliance, too, displays a pattern of deviations that is broadly north:south in its major cleavage (Figure 2.13). There are several small pockets of adjacent constituencies where its performance deviated markedly from the national norm, however; these include parts of Leeds and Teesside where it did relatively well (Teesside included two constituencies in Stockton

10 percent or more overpredicted

5 percent or more underpredicted

Figure 2.14 Percentage Deviations from the Actual Vote: Non-Voting Predicted from 1979 Vote and the National Flow-of-the-Vote Matrix

whose incumbents were founder-members of the SDP), as it did also in parts of south-east and south-west London and in Devon and Cornwall. In general, the Alliance performed badly – relative to its national performance – in the West Midlands and much of Greater Manchester.

Finally, the main implication to be drawn from the map (Figure 2.14) of non-voters is of an urban:rural split. The majority of Greater London constituencies recorded non-voting levels 5 per cent or more above those predicted, and there were similar concentrations in the inner urban constituencies of Greater Manchester, Merseyside, Tyne and Wear, the West Midlands and West Yorkshire. High levels of non-voting were also a feature of many retirement centres, such as Blackpool, Morecambe and Fylde, the Wirral and Bournemouth. The areas where turnout was substantially (10 per cent or more) under-predicted included much of the northern Home Counties, and the Southwest. (Inclusion of the latter indicates the lack of a simple correlation between high levels of 'non-voting' and population age.)

Just as the 1983 voting pattern by constituency could not be predicted successfully in most cases from the national cleavage, so it cannot be predicted accurately from the national flow-of-the-vote matrix. There was clear geographical variation in the pattern of inter-party shifts.

In Summary

This chapter has given unequivocal support to the thesis being developed in this book: knowledge of national patterns of party support and of partisan shifts is insufficient to provide accurate predictions of the result in each constituency. The conclusion must be that geography matters; it is not sufficient to know *what* you were, or how you voted in 1979; the analyst needs also to know *where* you were.

3 VOTES AND SEATS: THE GEOGRAPHY OF REPRESENTATION

The review of previous analyses of electoral behaviour in England (Chapter 1) indicated that the conventional wisdom prefers the 'class-equals-party' model and that, despite evidence to the contrary from aggregate data analyses, geographical variations in this equation are ignored. Those who reject the 'class-equals-party' model replace it with others in which geography is also considered unimportant. The analyses in Chapter 2 suggest that these alternative models are also insufficient representations of the behaviour of the English electorate. A 'one nation' perspective hides substantial geographical variations. The 'class-equals-party' model is a relatively poor predictor of the partisan distribution of voters in individual constituencies, as is the national flow-of-the-vote matrix.

Geographical variations in voting behaviour that cannot readily be accounted for by the conventional wisdom are undoubtedly real, therefore. *But are they important?* For the parties, importance relates to the allocation of seats. So, does the geography of votes affect the results of the election? If so, then additional weight will have been given to the thesis being advanced here regarding the necessity of a geographical perspective on electoral behaviour.

A brief inspection of the results of the 1983 election by region strongly suggests the need for a geographical perspective. Table 3.1 shows the distribution of seats between Conservative and Labour, by region, and also the percentages of seats where the Alliance came second to each of the others. In terms of representation, Labour is clearly the party of the northern regions plus, to a lesser extent, the industrial-urban centres of the West Midlands and Greater London. The Conservative Party dominates elsewhere (Figure 3.1), and in those areas the Alliance generally came second. The implications of 'two nations' are clear: in the urban-industrial regions, Labour is still strong, and Conservative is the second most popular; elsewhere, it is a virtual Conservative monopoly of the seats, with the Alliance in second place.

Table 3.1 The Regional Distribution of Seats, 1983

	Conservative	Labour	Alliance Second to Conservative	Alliance Second to Labour
North	22.2	72.2	37.5	26.9
York/Humberside	44.4	51.9	66.7	28.6
Northwest	49.5	51.7	52.8	5.9
West Midlands	62.1	37.9	52.8	0
East Midlands	80.9	19.1	50.0	0
East Anglia	90.0	5.0	72.2	0
Greater London	66.7	31.0	53.6	23.1
Southeast	98.1	0.9	84.0	0
Southwest	91.7	2.1	90.9	100.0

Percentage of Seats

Is this the pattern that the 'class-equals-party' model would predict, or would that have produced a different result for the 1983 general election? To answer this question, the survey analysis data have been used to suggest what the 1983 result would have been, if there were no geographical variations.

Occupation, Social Class, Tenure and the 1983 Result

If social class were the sole determinant of voting in England in 1983 and there were no geographical variations in the partisan choices of members of the various classes, the result would have been a complete Conservative whitewash: using the matrix in Table 2.3 to predict the result in each constituency, then every one would have been won by the Conservatives. Labour's voting strength in the manual classes was not spatially segregated enough to provide the party with any victories.

Using the combined occupation and tenure matrix (Table 2.3), the predictions are nearly as extreme. Labour would have won 23 seats – with the largest concentration in Greater London (Figure 3.2). Of those 23 it actually won 21, losing one in Nottingham to Conservative and one in Southwark to the Alliance. The other 500 seats would have recorded Conservative victories.

Clearly, the geography of voting was crucial in determining the allocation of seats; Labour did very much better in many areas than its national voting pattern suggested, winning seats that otherwise would have gone to the Conservatives.

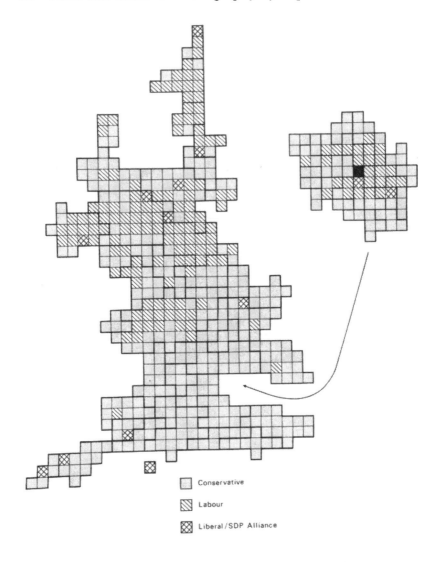

Figure 3.1 The Geography of Seats Won in 1983

The Flow of the Vote and the 1983 Result

Turning to changes between the 1979 and 1983 elections, the analysis in this section seeks to establish whether there were place-to-place variations in changing partisan behaviour that were substantial enough to affect the allocation of seats. Simple analyses suggest that there must have been. Table 3.2 lists the

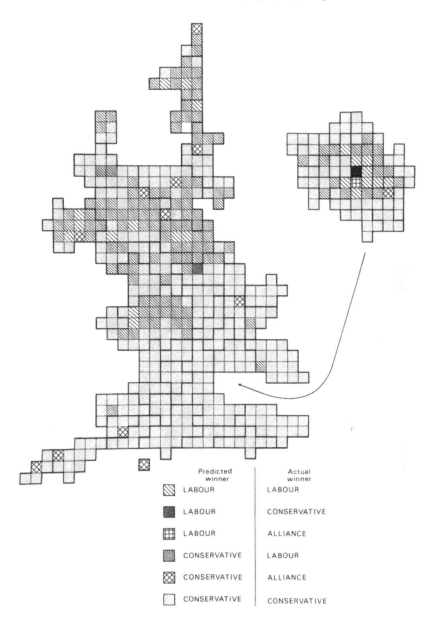

Figure 3.2 The Geography of Seats Won and Predictions from Occupation and Tenure

percentage change in each party's vote by type of constituency, using the Boundary Commission's classification within the context of the local government system; the Conservatives clearly lost most ground in the industrial-urban areas. Table 3.3 reports similar percentages according to the level of unemployment in each constituency at the 1981 Census; the greater the unemployment, the greater the Conservative loss of votes, the smaller the Labour loss and the greater the Alliance gain. Did these variations, and possibly others, affect which party won which seats, where?

Table 3.2 The Changing Pattern of Votes by Constituency Type

	Conservative	Labour	Alliance
Inner London	−16.1	−22.1	+122.7
Outer London	− 8.3	−34.5	+ 82.2
Metropolitan − Borough	−12.8	−18.3	+100.4
Metropolitan − County	− 7.1	−24.2	+ 82.2
Non-Metropolitan − Borough	−11.5	−25.8	+104.6
Non-Metropolitan − County	− 1.6	−31.3	+ 83.5

Table 3.3 The Changing Pattern of Votes by Level of Male Unemployment in 1981

	Conservative	Labour	Alliance
16.0%+	−14.3	−16.1	+135.3
12.0 − 16.0%	−11.1	−21.1	+132.3
8.0 − 11.9%	− 5.9	−25.5	+ 85.5
4.0 − 7.9%	− 2.4	−37.8	+ 60.6
0 − 3.9%	− 2.3	−47.2	+ 42.0

Using the predictions of the distribution of votes in each constituency, obtained by applying the matrix in Table 2.4B to the estimated 1979 result, a 'predicted' result for 9 June 1983 can be derived and compared to the actual result. This produces the matrix in Table 3.4.

One conclusion stands out from this table; if the national flow-of-the-vote had applied everywhere, the extent of the Conservative landslide would have been much greater, and Labour would have been left with a rump of only 75 seats in England. In almost exactly half of the seats that it won, therefore, the Labour Party did so because its loss of support to the other parties was less than the national pattern.

Table 3.4 The 1983 Result, and the Result Predicted from the National Flow-of-the-Vote Matrix

Actual Victor in 1983	Predicted Victor in 1983		
	Conservative	Labour	Alliance
Conservative	372	0	0
Labour	73	75	0
Alliance	7	3	3

Not surprisingly, given the spatial concentration of Labour victories (Figure 3.1), it was in the north – especially the urban-industrial north – and in the West Midlands and Greater London that Labour was able to retain support against the national trend and so bolster its position in Parliament. Figure 3.3 shows where the relevant constituencies are. In the south of England, Labour is virtually absent; if the national trends had applied in all constituencies there, it would have lost Ipswich and Thurrock and 60 per cent of its seats in Greater London, leaving it with just ten seats in London (mostly in the East End) and one in Bristol. In the West Midlands, too, it would have lost the majority of its seats.

Further north, the Labour hold on many of the seats was so strong that the national trend would have left them in the party's hands; this was especially so in Tyne and Wear, South Yorkshire, and the Merseyside-Greater Manchester boundary area. But many other seats were clearly at risk. Labour now has no representation in industrial centres such as Nottingham and Bury: the national trend would have excluded it from Warrington, Oldham, and the industrial areas of West Cumbria, too. As in 1931 and 1935, Labour could have become the party of the coalfields, Inner London, and little else (Johnston and Doorn-kamp, 1983).

In relative terms, the Alliance would have suffered even more than Labour if the 1979-83 national gross switching of votes had occurred in all constituencies. It would have held on to only three seats (Figure 3.4): Berwick, Rochdale and Truro. Three others (Leeds West, Southwark and Woolwich) would have been won by Labour, and the remaining seven by Conservative. Thus although, as Crewe's (1983a) discussion made clear, the Alliance was the net winner in terms of votes in 1983, in terms of seats it

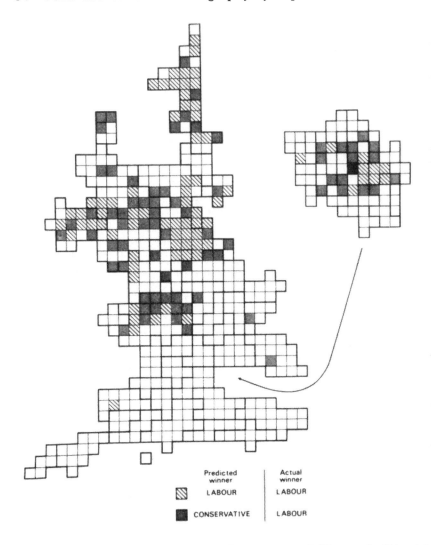

Figure 3.3 The Seats won by Labour, and Those it Would Have Yielded to the Conservative Party if the National Flow-of-the-Vote Matrix Had Applied There

would have been a net loser if there had not been substantial geographical variations in the strength of the various flows. The Liberal Party would have lost Colne Valley, Cambridgeshire NE (Isle of Ely), Isle of Wight, Liverpool Mossley Hill and Bermondsey and would have failed to win Yeovil and Leeds West; the SDP would have failed in Stockton South, Plymouth Devonport and Woolwich.

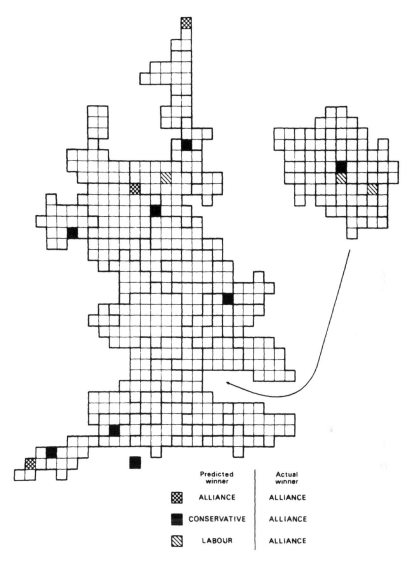

Predicted winner	Actual winner
ALLIANCE	ALLIANCE
CONSERVATIVE	ALLIANCE
LABOUR	ALLIANCE

Figure 3.4 The Seats Won by the Alliance, and Those it Would Have Yielded if the National Flow-of-the-Vote Matrix Had Applied There

Conclusions

In parliamentary terms, the 1983 general election was a Conservative landslide victory. The brief analyses here have suggested that, were it not for geographical variations in support for that party, variations that are not simply reflections in the composition of the electorate, the landslide would have been even greater. The Conservative Party could, if the conventional wisdom of many electoral analysts were correct, have won all but a handful of the 523 seats in England.

Why didn't it? This is the question addressed in the remainder of this book. What is clear is that the Labour Party is much stronger in most of northern England than one would expect, given the occupational and tenurial composition of the constituencies there, and that it is able to hold on to this support, against the national tide.The next chapter discusses reasons for this, and the remainder of the book tests the hypotheses formulated there.

4 ACCOUNTING FOR GEOGRAPHICAL VARIATIONS

The preceding chapters have demonstrated the existence of substantial geographical variations in voting behaviour at the 1983 general election in England, variations that cannot be accounted for solely in terms of the propensity of members of different groups within the population to support different parties and which in turn had a significant effect on the result of that election. Given that other analyses (e.g. Johnston, 1983a) have indicated continuity in the geographical pattern of voting within England over a long period of time, these findings pose a substantial question for electoral analysts: why do people in the same social position apparently differ in their voting behaviour according to where they live?

The general model of voting behaviour adopted in the literature suggests that there is no spatial variation in the processes of political socialisation. Parties exist to 'sell' a particular message to the electorate, each member of which is treated as an independent decision-maker. The probability that an elector will prefer a particular party depends entirely on that individual's position in the social structure. Place is irrelevant to such a model.

Such a stance is an excellent example of what Alker (1969) terms an *individualistic fallacy*, involving the separation of the individual from the context within which the voting decision is made. According to Key and Munger (1959, p. 281) such an approach, based on the belief that 'social characteristics determine political preferences', leads to a situation which

threatens to take the politics out of the study of electoral behavior

They argue the necessity of studying the individual *in context*.

There can be no doubt that there is at times a high degree of association between readily identifiable social characteristics and political preference . . . Yet there seems always to be a very considerable part of the electorate for which no readily isolable social characteristic "explains" political preference . . . Some of the considerable variance unaccounted for by social determination might be removed by attempts to analyze the nature of the individual's identification with the community and the nation (pp. 298-9)

At least part of the voting decision may be influenced (to claim that it is 'determined' seems too strong) not only by 'who you are' but also by 'where you are'.

The analyses so far in this book suggest that this conclusion almost certainly applied in England in 1983. The remainder of the book investigates possible reasons why this came about. The present chapter reviews the literature on this topic, presents an overview of the political situation in England in 1983, and outlines hypotheses relating the geography of voting to the economic and social geography of the country.

The 'Conventional Wisdom': The Neighbourhood Effect

The commonest explanation offered for the sorts of geographical variations described earlier in this book is generally termed the *neighbourhood effect* (other terms include 'contextual effect' and 'breakage effect': for a full review, see Taylor and Johnston, 1979; Rumley, 1979, 1981). This is a variant of the more general *structural effect*. According to this, individuals may be placed in situations in which their personal predispositions are at variance with those of others with whom they are in contact. Such contact sets up cross-pressures, and the greater these are the more likely that the individual will abandon personal predispositions, at least temporarily, and adopt those of his or her contacts. For many people, the local environment – or neighbourhood – is the source of many social contacts. The greater the explicit level of support for a certain political party in an area the more likely it is that residents who, according to their individual characteristics, would not be expected to support that party will be converted to the majority viewpoint. Thus the hypothesis of a neighbourhood effect suggests that the majority opinion is accentuated in any area: the greater the proportion of an area's electorate predisposed, by their class position, towards the Labour Party, the greater the proportion of others in the area who will also vote Labour.

Variants of this model have been tested many times, using both survey and aggregate data. (Most studies face problems relating to their data: the most sophisticated – e.g. Wright, 1977 – combine the two types.) Analyses tend to confirm the model, especially with regard to voters not strongly committed to a particular partisan preference. This suggests that local pressures to uniformity in voting behaviour do operate, especially with regard to politically-relevant issues, but the processes by which such uniformity is brought about are imperfectly understood (despite clear statements of the problems involved from the outset: Riecken, 1959; Luce, 1959). The remainder of this section illustrates the validity of that conclusion with respect to British studies that use the neighbourhood effect to account for spatial variations in voting behaviour.

The Pros

As indicated in Chapter 1, some British voting studies, especially but not exclusively those using aggregate data, have identified important place-to-place variations in party choice, holding constant social and economic cleavages. Thus, for example, Crewe and Payne (1976, pp. 79-80) conclude their discussion of variations in voting behaviour at the 1970 general election, which suggested, for example, that working-class electors are more likely to vote Labour in constituencies where the Labour Party is strong, by arguing that

> A process of political self-selection may be part of the explanation: strong Labour areas may be politically and socially uncongenial to workers voting Conservative, who therefore leave (or refuse to move in), whereas non-manual Labour voters may be peculiarly prone to find such areas attractive and therefore to stay put (or move in). It is also possible that the occupational composition of both manual workers and non-manual workers in strong Labour areas is skewed towards relatively low-status occupations, thereby producing an unusually high Labour vote. *But community influence is probably the strongest factor* . . . [although] the partisanship of manual workers is more open to the influence

of the locality than is that of non-manual workers (emphasis added)

A similar conclusion was reached by Butler and Stokes (1974, p. 129), from several different analyses:

> A good deal of empirical support can be found for the *principle* that once a partisan tendency becomes dominant in a local area processes of opinion formation will draw additional support to the party that is dominant (emphasis added)

At the regional scale, for example, they show that an overall difference of 8.9 per cent between the support given to Labour in the Northern and Southeastern regions can be decomposed into 2.3 per cent due to differences between the regions in their class composition, 5.7 per cent to differences in Labour support within classes between regions and 0.9 per cent to the interaction of these two causes (p. 126). At the local scale, too, they show that of those voters who identify themselves as working class, 50 per cent voted Labour in the constituencies with the largest middle-class populations, and 79 per cent in the constituencies with the largest working-class populations. Finally, using survey data referring to more than 120,000 respondents, they show that (a) the greater the middle-class proportion of a constituency's electorate, the greater the proportion of the middle class who vote Conservative, and (b) the greater the working-class proportion of a constituency's electorate the greater the proportion of the working class who vote Labour.

Like Crewe and Payne, Butler and Stokes (1974, p. 133) associate such variations with the neighbourhood effect hypothesis, operating in a variety of ways to exaggerate majority class opinion.

> Several processes can have helped to exaggerage a partisan tendency that had already been imported to the middle and working class areas by perceptions of class interest. Some writers have argued that a local electorate will perceive and conform to local political norms. Others have suggested that this tendency reflects the persuasive influences of informal contacts on the shop floor, in the public house and other face to face groups of the elector's world. These processes will

draw those who hold a minority opinion towards the view that is dominant in their local milieu, exaggerating the strength of a leading party and altering the pattern of party support by class.

And, like Crewe and Payne also, they fail to make those processes explicit.

More detailed discussion of the possible nature of such processes has been provided by Miller (1977, 1978). As noted in Chapter 1, his comparative analyses of survey and aggregate data strongly indicated the sort of exaggeration of the majority party vote in an area identified by Butler and Stokes. He accounts for this by suggesting that contact within a local area should stimulate political responses (Miller, 1977, p. 48)

> social contacts are structured by family, choice of friends, social characteristics and locality. If party appeals to group interest or group attitudes evoke any differential political responses, the patterns of contact between individuals will tend to increase the political consensus within high-contact groups.

Thus, where a party's appeal is directly to individual voters, according to their social position only, one will get the normal social cleavages. Where, in addition, the social contacts of individuals are structured according to their class position, that cleavage will be accentuated – the individuals' responses to the party arguments are reinforced by the responses of their peers. Most people, he noted, have most social contact with others similar to themselves, but some contacts are with dissimilar others. Drawing on the work of Putnam (1966), Segal and Meyer (1969) and others, he shows that the relationship between the partisan context of an area and the probability of contacts with members of the majority is S-shaped. Assuming two partisan groups, if they are of an equal size in an area there is no majority opinion to be exaggerated. As the relative size of one group increases, so the probability that a group of contacts has a majority from the dominant group increases – with the larger the group the greater the probability – up to a certain degree of polarisation. (In an area in which one group is predominant, the structure of an individual's contacts will almost certainly be the same as the structure of the population as a whole.) Thus the effect of the local environment is likely to be greatest in the area with between about 30 and 70 per cent of the population favouring one partisan position (Johnston, 1978); most, if not all, areas (other than those very restricted in extent) probably fall within this range.

Miller's analyses led him to conclude that (1977, p. 65)

> At a minimum, the class characteristics of the social environment have more effect on constituency partisanship than class differences themselves, perhaps much more.

and that

> The effect of the social environment may be explained by contact models: those who speak together vote together.

This may operate in different ways, however. Citing work by Przeworski and Soares (1971), he suggests (the diagrams are in Miller, 1978, p. 206) that there are three possible environmental effects: (1) a consensual environmental effect, whereby the majority opinion is accentuated among the members of both classes; (2) a reactive environmental effect, whereby contact with members of the other class hardens class consciousness so that, for example, middle-class residents of working-class areas are most likely to vote Conservative and working-class residents of middle-class areas are most likely to vote Labour; and (3) a combination of the two, whereby, for example, the middle class may show a reactive environmental effect whereas the working class display a consensual environmental effect. Miller's (1977) analysis of survey data showed that consensual effects among both classes were normal, except on one issue; the middle class showed a reactive response in their attitudes to trade unions, with the strongest anti-union feelings being expressed by those who lived in working-class environments. (For an argument against this view, expressing voting for a party as a percentage of the electorate rather than a percentage of those voting, see Taylor, 1973; he suggests that 'safe seat apathy', whereby supporters of the minority party in a constituency are less likely to vote than are supporters of the majority party, accounts for the polarisation effect.)

Miller (1977, 1978) completed his analyses by seeking the 'best-fit' equations for predicting support for the two main parties at the constituency scale. For the Conservative vote, the best-fit was with the percentage of the workforce who were employers or managers, rather than percentage in non-manual occupations, the variable more frequently employed. These employers and managers – the controllers in his terms – form one of the 'core classes' in society; the other core class, the 'anti-controllers', are not as readily identifiable in census data. Miller concludes that the pattern of voting is thus produced in the following way:

1. The basic cleavage in society, between Conservative and Labour partisanship, is between the core classes – the controllers and the anti-controllers.
2. A minority of the electorate belong to one of these core classes, and vote accordingly.
3. For the majority of the electorate, their class position is a function of their links to these core classes – the stronger and more numerous their links, the greater the probability that they will vote with the relevant core class.
4. These links are occupationally determined, reflecting not only past and present but also anticipated future occupations.
5. The links are structured by inter-personal contacts, in the home, at workplace, in voluntary organisations, and in the local environment.
6. The relevance of these links is best demonstrated with reference to the controller class.

Thus, the larger the controller class is in an area, the greater the possibility of links with it for members of other classes, and the greater the probability of Conservative voting. Miller tested this model using three occupational classes: A, employers, managers and professionals; B, other non-manual plus skilled manual workers; and C, semi- and unskilled manual workers. He found that for every one percentage point increase in the size of the core class, the propensity to vote Conservative increased: by 0.5 percentage points among class A; by 2.7 percentage points among class B; and by 1.9 percentage points among class C. Thus, if class C represents the anti-controllers, then (Miller, 1977, p. 65)

> All classes are more Conservative in environments with high densities of controllers, but . . . The most sensitive to the environment are those at the margin between controller and controlled irrespective of whether they are manuals or non-manuals . . . it is likely that both sets of core classes were themselves insensitive to the environment.

These findings were confirmed in a range of other analyses. Other variables introduced into the equations accounted for only a small proportion of polarisation, including percentage employed in agriculture (suggesting more pro-Conservative links in rural than in urban areas, perhaps a function of the relatively small size of the average establishment: see also Piepe, Prior and Box, 1969) and percentage unemployed.

Miller's analyses are of cross-sectional data, investigating voting patterns at particular times. Somewhat similar modelling has been used by Butler and Stokes (1974) in their analyses of changes over time. They identify a paradox in the phenomenon of uniform swing. The implication drawn from the observation that the shift, in percentage points, from one party to another is approximately the same between two elections in every constituency, is (pp. 140-1)

> the view that national political issues and events are of paramount importance for the voter and that his chances of moving from one party to another are virtually identical everywhere.

In other words, the national 'flow-of-the-vote' matrix applies in every constituency. But this cannot be so, for it would produce geographical variations in swing, not a uniform swing. If, for example, the net loss to Conservative was 10 per cent of its supporters, then in a constituency in which it had 70 per cent of the votes at the first election the swing against it would be 7 percentage points, whereas where it had 30 per cent it would only suffer a 3 percentage point swing. The result would be not a uniform swing but a reduction in polarisation, suggesting the irrelevance of the measure of swing (for illustrations see Johnston, 1981a; Johnston and Hay, 1982). To counter this tendency, and to produce the observed uniform swing, there must be forces that reduce the loss in the first constituency but accentuate it in the second:

> When the tide is running against his party, why should an elector in a hopeless seat be three or four times as likely to change his vote as a fellow-partisan in a safe seat? (Butler and Stokes, 1974, p. 142)

The answer, according to Butler and Stokes, is the neighbourhood effect. This operates (p. 143)

> to retard swings against the losing party in seats where the party had been strong and amplify swings against the party in seats where it had been weak.

And the processes involved are those of social contact, just as they are in the Miller model. It is the social ambience of a constituency, or component areas within it, and

> the persuasive nature of personal conversation . . . the tendency towards homogeneity of political opinion in seats where one of the parties is already dominant would reflect the fact that the persuasive contacts in such an area are mainly in one direction (p. 144).

Butler and Stokes tested this conception by classifying the voters in their survey according to whether their main sources of information about electoral issues were national (basically the mass media) or local; the latter are those who

> apparently get their information mainly from personal conversation and are therefore exposed to political stimuli that depend more on their local social milieu (pp. 146-7).

And indeed the evidence suggests that this produced variations in their behaviour (though see Bodman, 1983). Between 1964 and 1966 there was a national swing against the Conservatives of about 3.2 percentage points. Among locally-oriented voters, however, there was a swing of 8.5 per cent *away from* the Conservative Party where it was strong in 1964. The result is uniform swing, of about 3.2 percentage points everywhere.

The Cons

Miller's (1978, p. 281) conclusions are that 'environmental effects are strong', so that 'The major determinant of constituency class polarization is what we have termed the environmental effect of the local balance of core classes'. The consistency between model and research results is substantial, but nevertheless (p. 283)

> The major component [of polarization] comes from the power of the environment to structure social contacts *plus the empirical fact (and it is only empirical, not logical)* that contact across class boundaries makes a consensual impact on partisan choice.

In other words, Miller cannot be sure that he has correctly identified the processes involved. His analyses contain no material on inter-class contacts, and the bulk of literature on these suggests that they are relatively slight – and that when they occur, potentially contentious issues such as politics are avoided (Fitton, 1973).

Similarly, Butler and Stokes can only infer that the processes underlying uniform swing are inter-personal contacts.

It is this lack of a detailed analysis of the processes underlying the observed realities that led to Dunleavy's (1979) major criticism of the concept of the neighbourhood effect. He argues that analyses such as those quoted here are empiricist, and that although the role of class in the explanations for voting has been given a theoretical base, that of all other variables is treated in a 'remarkably untheoretical manner' (p. 411).

The role of personal contacts, of the sort implied by the neighbourhood effect, comes in for particularly scathing criticism. Residence in an area dominated by members of another social class is supposed to skew an individual's contacts towards that class, to promote the dominant class norms and to lead to emulation of the majority position, including its attitude to the political parties. But (Dunleavy, 1979, p. 413)

> Theoretically, inter-personal influence models have never explained which causal mechanisms affect political alignment, given that voting by secret ballot is hardly in the public

realm . . . and that political alignment is not apparently involved in any extensive way in the social life of the locality (unlike the work place). Empirically, these models have never been effectively connected with any evidence of the extensive community social interaction that is essential if they are to have plausibility. We cannot simply assume that political alignment brushes off on people by rubbing shoulders in the street, as exponents of 'contagion models' invariably seem to imply.

Such a pungent criticism would seem to have a great deal of force (it is expressed also by Prescott, 1972, 1978); the processes of interaction which are believed to underlie the neighbourhood effect do not appear to be very common (despite the modelling in Cox, 1969). And yet there is a substantial volume of literature that shows patterns of voting entirely consistent with the neighbourhood effect (Taylor and Johnston, 1979, ch. 5), even when Dunleavy's alternative model based on consumption sectors is incorporated (Garrahan, 1977; Johnston, 1983b, 1983c). Why?

Extending the Model

Studies that argue the relevance of the neighbourhood effect postulate a general process of 'voting influenced by contagion' that is operating in all places. The argument presented here (based on Johnston, 1984a) is that this is but one of the possible influences leading to geographical variations in voting – and possibly only a minor influence. Other processes may well be in operation, and their nature is outlined here.

People vote for a particular party at an election for a variety of reasons. Some may do so out of habit; they have traditionally voted for that party, having decided that it best represents their interests, and repeat the action with little further consideration. Others may do so because of their wish to support an individual within that party, usually either the local candidate or the party leader (who will head the government if the party is successful overall). Some may do so because of the party's stand on a particular issue, which to them is the most salient of those on which the election is being fought. Others may do so because on balance, considering the parties' positions on several salient issues, one party seems best able to represent their interests. And, of course, some may vote for a particular party or candidate because their friends and neighbours suggest that they should. The result is expressed very generally by Berrington (1983, p. 267) regarding differences in swing at the 1983 election:

> These differences almost certainly reflect both tactical voting . . . and in addition the pressures from the social milieu, working at the local and constituency level, which tend to weaken the support of the smaller of the two major parties.

Whatever the dominant process of decision-making, it is possible that the voter's environmental context(s) will be influential – though not necessarily through inter-personal contacts. As Schattschneider (1960) suggests for the United States, in democracies such as Britain's the electorate is only a 'semi-sovereign people' because the agenda at an election is structured by others – the political parties and the media. Thus the issues to which electors respond are those which are presented to them: the parties, or some of them, may misread the electorate's interests and fight on the wrong issues, as happened to the Heath government in Britain in 1974 and the Whitlam government in Australia in 1977. But the parties are frequently very sophisticated in their choice of agenda items (see Pool *et al.*, 1966, on the strategy developed by Kennedy's campaign team in 1960).

How voters respond to the agenda items may reflect contemporary reactions only, but it may be influenced by attitudes to those issues developed in earlier periods of political socialisation. For example, in the United States Presidential Election of 1968 George Wallace put racial segregation on the agenda. As Wright (1977) shows in an analysis of voting for Wallace, white voters in the southern States were more likely to vote for Wallace the greater the size of the black population in their home county in 1940, a date chosen because 'it predates the Civil Rights Movement and catches much of the adult population in the formative years of adolescence and early adulthood' (p. 500: the correlation between voting for Wallace and blacks as a percentage of the population in the home county in 1940 was 0.38; for blacks in the home county in 1970 it was 0.33; holding 1970 constant, the partial correlation with the 1940 percentage was 0.2

whereas holding 1940 constant the partial correlation with 1970 was 0.001).

This greater support for Wallace among people socialised in areas with large percentages of blacks could be the result of one or both of two processes, according to Wright. The first is the classic neighbourhood effect: 'context influences behavior through processes of interpersonal influence', especially for those active in local social organisations and who are therefore open to peer-group interaction and influence. The second mechanism excludes inter-personal contact and influence: context influences attitudes, so that increased exposure to blacks increases racial hostility. Wright's analyses favoured the former:

> context does not influence behavior directly but does so through intervening social and attitudinal processes (p. 507)

The context that the peer-group interprets for the voter comprises several spatial scales; not only is it the concentration of blacks in the home county that influences racial hostility, but also their concentration in the home State and whether that State is in the Deep South; in such States, racial issues were more salient politically, and more likely to influence white voters.

According to this example, the political socialisation of voters in a particular environmental context provided the basis for later voting decisions, when that context again became relevant to the political agenda. In other situations, it may be that once a party or candidate has exploited an environmental context the strength that is developed there leads to continued support long after the original stimulus has become irrelevant. This was clearly demonstrated by Key and Munger (1959) for the case of Southern Indiana, an area of continued Democrat strength since the end of the Civil War. This block of counties was largely settled, after 1868, by people from the southern States, among whom the Democrat Party had gained almost total support because of its policies regarding white supremacy; further north, a block of strongly Republican counties reflected their settlement by Germans with strong anti-slavery sentiments. Key and Munger argue that whereas these post-Civil War issues may have been salient until the end of the century they were much less potent after. And yet the pattern of voting established by 1868 persisted right through until 1948 (the last year of their study). Once a party has established hegemony in an area, they suggest, then it tends to retain voter support, long after the original reason for that hegemony has disappeared, because (Key and Munger, 1959, pp. 286-7)

> the dominant classes of the community allied themselves with the party whose policies of the moment were most akin to their inclination. Doubtless great contests and stirring events intensified and renewed partisan loyalties. The clustering of interests, career lines, and community sentiments about the dominant party gives it a powerful capacity for survival.

Thus local political socialisation, for new generations of voters, takes place in an environmental context where the continued strength of a party – reflecting its relevance to a crucial issue in the past – overrides more general (extra-local) considerations.

This Southern Indiana example reflects the mobilisation of the electorate by a particular party in a pre-political vacuum, when loyalties have not been established there. (A further example of such mobilisation of a periphery is provided by Burghardt, 1963, 1964.) It may be that the mobilisation reflects trends elsewhere, with class cleavages developing in the new area that mirror those in other parts of the country. Alternatively, a particular, local pattern may be established, reflecting the salience of local issues over national ones. (See Cox, 1970a, on political mobilisation in urban-industrial and rural Wales: in the former, the English class-cleavage was introduced; in the latter, a protest movement – focused on the Liberal Party – against English landowners formed the basis for local political divisions.)

A major example of this type of local mobilisation is provided by working-class voting patterns in industrial Lancashire. Clarke (1971) argues that the conventional wisdom regarding political developments at the end of the nineteenth century is that Labour replaced Liberal because of its clearer identification with working-class interests and its better appreciation of the role of the state in a modern economy. But this does not fit the situation in Lancashire. Clarke argues that voting preferences developed in the context of antecedent social attitudes, and at that time

> while sometimes a clash of economic interests underlay political conflict, more often it did not . . . the importance of

religion – not in any narrow sense, and not in a spiritual sense, but *religion in a social sense* – should be stressed. For this was a religious age (p. 16)

Working-class voters in Lancashire, especially in the cotton towns, were mobilised by the Conservative Party, building on the local Anglican religious organisation (and in part against the link between Irish Roman Catholics in the area and the Liberals). And then, in the first decade of the century there was a major shift, precipitated by economic circumstances, and the Liberal Party mobilised the working-class vote. (See also Hobson, 1968, for a discussion of the way parties moulded political opinion in the 1900s, so that the working class voted Liberal and the middle class Conservative; Taylor, 1984b.)

In all of these examples, the general theme that stands out is that the electorate in any area is mobilised by parties seeking support. The basis of that mobilisation may be the same as in many other areas (i.e. the class-cleavage), or it may have reflected particular local issues and circumstances at the time. Once support for a party has been mobilised, however, it tends to persist. Several reasons can be suggested for this. The first is that once a strong attachment to a party has been formed by a voter, it is difficult to break – what Butler and Stokes (1974, p. 58) call 'The hardening of partisanship in the mature elector'; their data show that the older the voter the smaller the probability of changing allegiance between elections. Thus an established hegemony in an area becomes built-in to the local political culture. Secondly, the major influence on political socialisation is the home and family, so that partisan attitudes are transmitted generationally; again, this is illustrated by Butler and Stokes' (1974, p. 51) data

A child is very likely indeed to share the parents' party preference. Partisanship over the individual's lifetime has some of the quality of photographic reproduction that deteriorates with time: it is a fairly sharp copy of the parents' original at the beginning of political awareness, but over the years it becomes somewhat blurred, although remaining easily recognizable.

Thus, once a party has mobilised support in an area, this is carried forward. Thirdly, the successful party is likely to have a strong local organisation, and to control local government, giving it great visibility.

Two conclusions stem from this. The first is that continuity in voting patterns over time is to be expected; where a party is strong at one election it is likely to be strong (in relative terms at least) at the next, and at the next after that . . . The second is that once a party has established its strength in an area its greater control of resources there, with which to mobilise votes in the future, should enable it to retain above-average support. American electoral studies show the immense advantage of incumbency there, because only one individual candidate can be associated with successfully representing the area's interests (see Johnston, 1980). Individual MPs have less power over the purse in Britain, but a party that holds a constituency does have considerable advantages – exposure in the media for its MP, for example, a strong local organisation and ability to raise campaign funds (incumbent parties tend to spend more: Johnston, 1979a), and perhaps links to a similarly-oriented local government. It becomes 'the party that represents the people of this area'. It may undoubtedly stimulate some 'reactive environmental' effects (to use Miller's term: see p. 62) but in general it may mean that the combination of local socialisation and established local presence ensures greater support for a party from all groups in a community than it would get where it is weak, support which results in the shifts of preferences that produce the uniform swing.

Occasionally, that support may be substantially eroded. An important local issue arises, and one of the parties mobilises support around it, thereby altering the balance of voting. (Curtice and Steed, 1982, suggest, for example, that the flow of votes away from Labour in 1979 was less than average in the fishing ports, because of Labour's stance on the EEC fishing policy.) The issue may be local, in both origin and political response. Or it may be national in response, with one party seeing an issue as likely to win it votes in areas where it is salient. Thus the geography of voting may be shifted, perhaps substantially and permanently (see Archer and Taylor, 1981; Johnston, O'Neill and Taylor, 1983) or perhaps only for one or two elections in one or two places (see Taylor and Johnston, 1979, pp. 294ff. on Enochland).

Even with the latter, a shift in traditional allegiances at one moment becomes partly embedded in the local political culture. A political party may be able to mobilise support in an area where it has traditionally been weak – perhaps at a by-election as a protest against an unpopular incumbent government. The temporary shift that it produces may not be immediately righted. For example, the mining constituency of Ashfield, Nottingham-shire, was traditionally a Labour stronghold; from 1955 to 1970 Labour only once failed to win more than 70 per cent of the votes cast, and in the two 1974 elections it won about 60 per cent. And then at a by-election in 1977 Labour lost the seat, and a majority of 23,000 votes. It regained it in 1979 at the general election, but with a majority of only 8,000 and only slowly returned to its former dominance (its majority in 1983 was 16,000).

The tendency for a party to 'invade' a constituency at a by-election and score a notable success, partly as a protest against an incumbent government (and perhaps the opposition too) and partly because of its ability to mobilise voters around a local issue, has been typical of the Liberal Party in England since *c.* 1960. The Isle of Ely constituency, for example, had no Liberal tradition (between 1950 and 1970 the party only once stood at a general election there – winning 11.4 per cent of the votes in 1966). At a by-election in July 1973, the Liberal Party won the seat, with a majority of 3.3 percentage points. Its majority increased in February 1974 to nearly 15 per cent, remained at that level in October 1974, was cut to 5.9 per cent in 1979, but increased again to 9.8 per cent in 1983.

The strength of a party's position in an area and its political culture may be either enhanced or reduced by other trends. The population of an area changes, for example; in the long term there may be substantial growth or decline, whereas in the short term there is almost certainly a large amount of movement producing little net change. With regard to the latter it is generally assumed that the balance, in terms of political attitudes, between in- and out-migrants is roughly equal. Nevertheless, if an area has a strong political culture favouring one party, and obtains more support from certain social groups than is the case nationally, then migration should dilute its support unless one or more of the following holds: the out-migrants are those less likely to conform to the local culture; the in-migrants are from areas with similar cultures; the in-migrants are self-selected in terms of a predilection for the local culture to which they are moving; or the in-migrants are readily socialised into the local culture, via contagious processes. All have been suggested as reasons for observed voting, with varying degrees of plausibility; one might accept, for example, that middle-class in-migrants to a coal-mining area are more likely to have working-class antecedents and to vote Labour than are middle-class in-migrants to an exclusive suburb, but might doubt hypotheses regarding migration and contagion (Cox, 1970b).

Over the long term, Taylor (1979) has provided strong, though not conclusive, evidence that changing party fortunes in Britain are associated with migration; the areas that became more pro-Labour between 1964 and 1979 were those that suffered population decline, with the out-migration probably dominated by pro-Conservative individuals. Four types of voter can be suggested (Johnston, Hay and Taylor, 1982): loyal stayers, whose voting behaviour and place of residence do not change between elections; loyal movers, who migrate but vote for the same party; switching stayers, who don't move but who change allegiance; and switching movers, who change both residence and party supported. A local political culture is best maintained by a hard core of loyal stayers; in a society characterised by increased social and geographical mobility, their relative numbers may decline, and the continuity of the geography of voting would decline accordingly.

Somewhat similar arguments have been put forward in a seminal paper by Curtice and Steed (1982). They argue that two major patterns of geographical variation characterised trends in UK voting behaviour during the period 1955 to 1979: a north:south cleavage and an urban:rural cleavage. The north of England, Scotland and the urban areas have swung to Labour, the south of England, the Midlands and the rural constituencies to Conservative. (This trend was apparently replaced in 1983 – see Chapter 2 – by the popularity of the Alliance in the latter group of constituencies, relative to Labour, but not in the former group.) This, they claim, was a new trend; before 1955, there was no significant regional or urban:rural voting cleavage. Furthermore, their data suggest that the rate of growth of these new cleavages has accelerated, being faster in the 1970 to 1979 period than between 1955 and 1970. (Michael Steed, in a personal communication, has suggested that the trend increased yet again in the 1979 to 1983 inter-electoral period.) The reason for this, they suggest (with Taylor), is differential migration. There is a changing distribution of the 'core classes' (to use Miller's term again), with a growing concentration of the 'controllers' in the south and in the rural constituencies, countered by a growing concentration of the 'anti-controllers' in the north and the urban areas. The foundations for sectional effects are being strengthened.

Other analyses confirm the existence of a polarisation effect between each pair of general elections, with each party in relative terms doing better than expected where it was relatively strong but worse than expected where it was relatively weak (relative, that is, to its share of the vote at the first of the two elections: Johnston, 1981b, 1981d). However, analyses of a sequence of inter-election periods (Johnston, 1981a) from 1966 to 1979 provide no clear evidence of growing regional polarisation.

Towards a Typology of Local Effects

The previous discussion has charted a middle course between those, such as Miller, who promote the contagion model of the neighbourhood effect as the sole cause of geographical variations in the strength of an electoral cleavage and those, such as Dunleavy, who reject that model entirely. There is evidence that contagion does operate, as in Wright's study of the Wallace vote. But it is unlikely that it is general, and it is more likely to be relevant in certain contexts, such as a salient issue like race relations, than in others. What is important is that it is not necessary to rely on the contagion model to account for the many observed geographies of voting that are consistent with it.

It is proposed here that there are four types of effect that operate, singly or in combination, to produce the patterns of voting under discussion.

1. Contagious Effects. These result from inter-personal influence. It is assumed: (a) that people choose those with whom they discuss politics from within the local area; (b) that the greater the strength of opinion towards a particular voting decision that is encountered in those discussions, the greater the probability that a waverer will be converted to that opinion; and (c) that the choice of discussants will probably produce a majority who support the majority view in the area. Thus, the greater the level of support for a party in an area, the greater its pull on those who would otherwise vote for another party.

The extent of the local area within which such contagious processes operate is unspecified. It is almost certainly small, much smaller than a British constituency. For testing purposes with aggregate data this is irrelevant, however. If one assumes that most residential areas are dominated by one class, then if a constituency comprises 70 per cent in one class it is likely that approximately 70 per cent of its residential areas are dominated by that class. Thus, the existence of this effect or, at least, of voting patterns consistent with this effect, can be explored at any spatial scale of data availability.

2. *Sectional Effects*. These represent the political culture of an area, whereby voter attitudes have been mobilised and manipulated by political parties and other agents so that a particular partisan position dominates there. This dominant position may be peculiar to that area – an area of Conservative Party strength where the social structure might predict Labour dominance, for example – or it may be an accentuation of the general pattern predicted from national data relating to electoral cleavages. Whichever it is, and the latter is by far the most likely, the sectional effect is long-lasting; it is deeply engraved within the local culture.

A consequence of the existence of a sectional effect, as discussed above (p. 69), is that the process of political socialisation in the home and in the various niches of the local environment is biased towards the strong party. That party is also likely to have considerable general influence in the local community, notably its local governments; it occupies a quasi-institutional position within the local social structure.

3. *Environmental Effects*. These differ from the sectional effects in being much shorter in their duration – at least up to the time of the election(s) being studied. The area is mobilised to vote in a certain way, at a particular election or sequence of elections. The shift from its traditional allegiance (its normal vote: Converse, 1966) may be temporary, and there is a rapid return to the previous situation. It may, however, presage a permanent shift, which lasts much longer than the original cause that stimulated it.

Two subtypes can be identified. The first are *candidate environmental effects*, in which the shift is a consequence of some characteristic of the candidate. The classic example of this is the 'friends and neighbours' effect, whereby voting decisions are made to favour a local candidate, with some abandonment of usual allegiances in order to vote in this way (Taylor and Johnston, 1979, ch. 6; in some countries, the parties employ this effect to mobilise the electorate: Parker, 1982). Voting is thus personalised; a change in candidate should lead to a return to the traditional cleavage, unless the strength of the original candidate is such that part of the personal vote is 'inherited' by a successor.

The second subtype is the *issue environmental effect*. A particular issue relating only to the local environment becomes the centre of political debate, and the stance of parties and candidates on this can lead to a temporary abandonment of traditional allegiances in order to vote on that issue alone. The racial issue exploited by the candidature of George Wallace in 1968 and Strom Thurmond in 1948 mobilised support on that particular aspect, mainly attracting votes from the Democrat Party; at subsequent elections, the traditional pattern of voting was re-established (Johnston, 1982a). The same happened with voting in the racial issue in Britain in the 1960s. But the traditional pattern may not be re-established. Instead, a new pattern is inaugurated, creating a new sectional effect. (The voting shifts in Northern Ireland during the last two decades illustrate this.)

Many issues could be the source of an *issue environmental effect*, hence the interest in pork barrel politics (Johnston, 1980). In Britain, geographical variations in the level of, and changes in the level of, unemployment could be the source of spatial variations in party choice, for example (Johnston, 1983d).

4. *Campaign Effects*. In most countries, the brief campaign period before an election involves the parties in (a) identifying where their supporters are; and (b) encouraging them to vote. Given the general efficacy of this operation, then the more active the campaign, the greater the returns should be in terms of the overall level of turnout and of the distribution of votes between the contestants.

Parties and their representatives will be most active as campaigners in locations where they think a high turnout among their supporters is most important. In the British system, this is in the marginal seats, those which could change hands given a small shift in the distribution of votes among two or more parties.

There, the parties will be active in encouraging their supporters not to defect, either to another party or to abstentions. The argument may well emphasise the dangers of a 'protest vote' for a third party; in a marginal situation between two parties, the party that loses most votes to a third party may well lose the seat and voters will be encouraged to ensure that this does not occur.

The resources used in a campaign are manpower (much of it free) and money, with the latter being employed to promote the party and candidate image via various forms of advertising. In the USA where large sums of money are invested in such advertising, mainly in TV, the returns on spending are great (see Caldeira and Patterson, 1982); the more that a party spends relative to its competitors, the more votes it tends to get. In Britain, where the sums spent are much smaller, there is little evidence that they are very efficacious (Johnston, 1983e).

Local Effects and the 1983 Elections

Two extreme positions can be taken regarding the geography of voting. The first is that people vote according to their situation with regard to the basic political cleavage(s) within society, and that where they live has no influence on their voting decision. The geography of voting is thus no more than a reflection of the geography of the political cleavage(s). The second position, at the other extreme, is that there is no national political cleavage, but rather a series of sectional, or regional, cleavages, independent of each other. According to this, where one lives is the prime determinant of how one votes, and understanding the geography of voting is crucial to understanding election results. The middle course, adopted here, is that there are national cleavages but these are modified locally, with the modifications being substantial enough to have a significant influence on the election result.

Regarding the national (i.e. English) cleavages, two are presented in the analyses that follow; occupational class and housing tenure. Occupational class represents the traditional political cleavage within British society. Despite the evidence of Sarlvik and Crewe (1983) regarding a 'decade of dealignment' in the 1970s and of Rose (1982a) on the contemporary invalidity of the 'class-equals-party' model, occupational class is still a major correlate of, and apparent influence on, voting behaviour, reflecting the traditional positions of the Conservative and Labour Parties. The latter, in particular, continues to maintain strong links with organised labour and sections of the party still argue strongly for identification with the needs and aspirations of the working class.

The second cleavage introduced here, housing tenure, reflects the importance of what Dunleavy (1979) terms consumption sectors, both to voters and to the political parties. Although the Labour Party is not committed to council housing as the sole housing tenure, it has promoted that tenure much more than the Conservative Party in recent decades and Labour is clearly identified by voters as the most committed party to a strong state housing sector as the best means of ensuring good standards of housing for the lower paid. The Conservative Party, on the other hand, is clearly identified with a policy of promoting owner-occupation, as part of its free market ideology and as a means of legitimating capitalist principles by co-option. It is also vigorous in promoting the sale of council houses to sitting tenants, the advance of its general policy goals and a reduction in the size of the public sector.

Other indices of the same cleavages could be introduced, as could further cleavages (such as those suggested by Rose, 1982a, and Franklin, 1982). They are not, however, because to multiply the number of cleavages studied would confuse the issues to be considered regarding the geography of voting. Furthermore, it would be difficult, if not impossible, to introduce many others because they are not represented in census data, on which the analyses rely heavily. Occupational class and housing tenure are, fortunately, not only the basic cleavages in the English electorate (the third most important is membership of a trade union) but also fully documented in the census collections. Parents' occupation, trade union membership and so on are not in the census, and desirable though it may have been to incorporate them in the analyses they had to be excluded.

There are other sources of voting differences that are represented in the census, but are excluded here. They include age, sex and race. In general terms, survey data suggest little variation between age groups in their political preference (Crewe, 1983a). Similarly, the differences between the sexes are slight, although these are little explored in detail – with regard, for example, to the labour force participation of females (Peake, 1984). It may be that there are substantial geographical variations in voting by age and sex which cross the traditional cleavages, but these are not explored here. With regard to race, claims are made both that the support of certain national groups for a particular party (usually Labour) is overwhelming and crucial, and also that the presence of certain national groups in an area influences how others vote. Again, these are valuable hypotheses worthy of detailed study, but they are excluded here.

Given that occupational class and housing tenure provided the basic categorisation of political cleavages in England at the time of the 1983 general election, then it is accepted that the class/tenure category occupied by an individual is the basic influence on voting. Following this, it is hypothesised that there are geographical variations in the propensity of individuals in a particular category to vote for the 'preferred' party: *people in the same category will have different political attitudes and will display different voting behaviour, depending on the environment in which they live.*

This hypothesis of local effects could be interpreted in many ways. Here, the basic interpretation reflects the assumed operation of *sectional effects*, in two ways.

1. The greater the proportion of an electorate who are in one of the 'core categories' relating to the support of a particular political party, the greater the support that the party will draw from all categories. In general terms, the Conservative core comprises white-collar residents of owner-occupied housing whereas the Labour core comprises blue-collar residents of council housing. (More detailed categorisations are explored in Chapter 6.) Thus, for example, the greater the proportion of white-collar residents of owner-occupied housing in a constituency, the greater the Conservative vote should be, from all categories, reflecting the local strength of the pro-Conservative ideology.

2. The greater the strength of a party in an area, the less likely it is to lose support at a period of general decline in the party's fortunes and the more likely it is to win support in a period of growth.

Together, these two arguments suggest that the electoral map of

England comprises a series of party heartlands in each of which, through a continuing process of political socialisation and activity, one party dominates. Relative to the country as a whole, the greater a party's basic strength (in terms of the national cleavages) in an area, the greater is its ability to win support from all categories of voter. Furthermore, this electoral map is extremely stable, so that major shifts are unlikely; over time, the parties retain their strength in their heartlands. (The process by which that strength is created and maintained is not specified. At least part of it may be due to contagious effects, but this cannot be analysed with the data available, and is in any case irrelevant to the general argument being advanced.)

These sectional effects are long-lived. At any election they may be 'invaded', if not overridden, by environmental and campaign effects, or by a new, national cleavage. The possibility of some of these local effects being present cannot be specified before the event in any detail – the existence of a personal vote for a particular candidate, for example – but can only be used to account (tentatively) for unexpected results in any analysis. Others can be hypothesised and tested for, such as above-average support for incumbents. In England in 1983, for example, unemployment was clearly a major issue on which political debate focused. It was also geographically variable. In areas where it was high, it could be anticipated that there would be a greater reaction against the incumbent government (which had presided over an increase in the level of unemployment from 5.2 per cent in June 1979 to 12.7 per cent in March 1983, even if, as it claimed, its policies were not responsible for that increase). Not only the unemployed but also those living in areas of high unemployment, and aware of the problems, might express their displeasure at this by voting against the Conservative Party, using this issue to override traditional cleavages.

Regarding campaign effects, the most important in 1983, as in previous elections, was undoubtedly constituency marginality. In general terms, the parties offered advice to their traditional supporters not to desert them, and certainly not to cast a protest vote for a third party (i.e. the Alliance) because this could 'let in' the opposition. In marginal constituencies, this message was pressed even harder: high turnouts and rates of loyalty were sought, and the potential threat of 'tactical voting' made clear.

Other effects, of less overall importance, undoubtedly

occurred, reflecting detailed sectional effects. For example, in rural areas the low levels of labour organisation and the greater deference of the working class, many of whom work in small enterprises, are generally thought to produce a greater propensity to vote Conservative among the working class. Conversely, in many areas, working-class solidarity is thought to produce above-average support for Labour. Finally, and more generally, there may be regional effects that provide a broader social context within which voting decisions are made. For example, the general hypothesis of a sectional effect suggests that above-average Conservative voting among the working class might be expected in a strongly middle-class constituency. The wider environment of that constituency may be very pro-Labour, however, and may influence the propensity of working-class Conservatism; an urban constituency in the industrial north might display less working-class Conservatism, *ceteris paribus*, than a rural constituency in East Anglia.

Conclusions

This chapter has reviewed the analyses and interpretations of local effects in voting patterns. It has been shown that such effects are common, and that they are usually associated by analysts with a hypothesised contagious process. The present review does not doubt the possibility of contagious processes operating, but doubts that they are either widespread or the major cause of local effects. Instead, it suggests a more general category of sectional effects, occasionally invaded and sometimes removed by environmental and/or campaign effects. These sectional effects ensure that the distribution of votes is spatially more polarised than is the distribution of members of the basic political cleavages in society. The greater a party's base of support in an area, the greater the support it can draw from all social groups, and the less likely it is to lose support when the national swing is against it.

These sectional effects operate at the local scale, and are aggregated up to the constituency level – the basic reporting unit for electoral data in England. Many constituencies are part of a wider context, however – a conurbation or region – and their location in this may also contribute to the sectional effect.

At the 1983 election, the major environmental effect in operation related to the economic policies of the incumbent government, especially with their impact on the level of unemployment. The greater the level of unemployment the greater the probability of a vote against the government. In addition, the government's promotion of working-class owner-occupation of housing could have stimulated substantial support in those areas where it produced embourgeoisement. Finally, there were the usual potential campaign effects, related to the marginality of constituencies and popularity of incumbent candidates.

Apart from the suggested environmental effects, these general propositions refer to any recent general election in England. But 1983 contained a new element, a third political grouping (the Liberal/SDP Alliance) with the potential to replace the Labour Party as the second largest in the country, if not to win the election. The Alliance projected itself as a radical new element in English politics, entirely apart from the traditional cleavages, and drawing equally from all groups within society. If such a party were beginning from scratch, it would presumably display no sectional effect, in that it was presenting itself as a national party. However, it had two foundations on which to build: the existing spatial pattern of strength for the Liberal Party, which provided the organisational core at constituency level; and the remnants of support for the initial SDP MPs, all but one of whom were defectors from the Labour Party whip. (Two other SDP MPs were elected during the previous Parliament, at by-elections; only one was for an English constituency – Crosby.) Thus one would anticipate a sectional effect in Alliance support in 1983, reflecting not variations in the distributions of various social groups but rather the earlier strength of the Alliance Parties.

The hypotheses to be tested here, therefore, are:

1. The greater the percentage of a constituency's electorate in the core class of Conservative supporters, the greater the support for the Conservative Party, from all classes. (In this, and other hypotheses, class is used to cover both occupational class and housing tenure.)

2. The greater the percentage of a constituency's electorate in the core class of Labour supporters, the greater the support for the Labour Party, from all classes.

3. The greater the support for the Liberal Party in a constituency in 1979, the greater the support for the Alliance Parties in 1983, from all classes.

4. Holding constant the relationships set out in the first three hypotheses, there should be regional and constituency type (metropolitan:non-metropolitan; borough:county; Inner London: Outer London) variations in support for the parties.

5. The greater the level of unemployment in a constituency, the lower the vote for the Conservative Party, among all classes, and the greater the vote for Labour and the Alliance.

6. The greater the percentage of working-class residents in a constituency who are owner-occupiers, the greater the support for the Conservative party among the working class and the lower the vote for Labour and the Alliance.

7. The greater the percentage of the workforce in a constituency in agricultural occupations, the greater the vote for Conservative, among all classes, and the greater the percentage in mining and related occupations, the greater the vote for Labour.

8. The greater the vote for a party in a constituency in 1979, the greater the loyalty of its supporters in 1983 and the greater its (relative) ability to attract voters to shift from other parties.

9. The greater the marginality of a constituency, the greater the loyalty of voters between the 1979 and 1983 elections.

10. Where incumbent MPs stood in 1983, the greater the vote for their party, among all social classes, and the greater the loyalty to the party between the 1979 and 1983 elections.

11. Where Liberal candidates stood in 1983, the greater the Alliance vote and the greater the loyalty to the Alliance from 1979 to 1983, compared to constituencies where SDP candidates stood.

12. Holding constant all the above relationships, the more that a party spent in a constituency and the less that its opponents spent during the campaign, the greater the loyalty to that party between 1979 and 1983 and the greater its ability to attract those who voted otherwise in 1979.

Testing these hypotheses occupies the remainder of the book.

5 TOWARDS ANALYSES: ESTIMATING THE DEPENDENT VARIABLES

The hypotheses outlined in the previous chapter suggest tests using the conventional regression framework, with the dependent variables referring to aspects of the geography of voting. However, most of the hypotheses suggest relationships that cannot be identified using the aggregate data available from the census plus the voting returns. For example hypothesis 1 suggests spatial variations in the percentage of members of a particular class group voting Conservative, but data with which to investigate such variations are not available. How, then, can the hypotheses be tested?

Three ways can be suggested:

1. Modifications to the hypotheses, so that survey data can be used to create the dependent variable (as in Wright, 1977). One could then test, for example, whether the probability of a white-collar person voting Conservative varied according to locational variables. Such an approach has much to recommend it, except for the problems of obtaining a representative sample of locational contexts – given that most surveys seek to represent the *national* electorate.

2. Use of the traditional regression approach, using the results of the regression equations to derive the needed dependent variables (as in Crewe and Payne, 1976). This requires a fully-specified regression equation (see Bogue and Bogue, 1982).

3. Use of a recently-developed procedure for estimating the values of the dependent variables in each constituency (as argued by Johnston and Hay, 1983; Johnston, 1983b).

All three have their benefits and disadvantages. The first is not considered here because of the problems with survey data relating to spatial variations. The third is adopted, but first the reasons for rejecting the second are outlined.

The Traditional Regression Approach

The dependent variables outlined in the previous chapter are of two types: those which relate to the percentage of people in a particular social category who vote for a specified political party; and those which relate to the percentage of people who voted for a specified party in 1979 voting for another (including the same) specified party in 1983. It is possible, under certain circumstances, to estimate these percentages via a regression model, incorporating other independent variables.

Crewe and Payne (1976) demonstrated this in their analysis of the 1970 general election. Assume an electorate divided into two social classes, with everybody voting for one of two parties only. This gives the table for a constituency

	Manual	Non-manual	
Labour	p	r	L
Conservative	q	s	C
	M	N	

where

L is the proportion of the votes won by Labour;
C is the proportion of the votes won by Conservative;
M is the proportion of the voters in manual classes; and
N is the proportion of the voters in non-manual classes.

Further

p and q are proportions of M, such that $p + q = 1.0$;
r and s are proportions of N, such that $r + s = 1.0$.
(Thus p is the proportion of manual workers voting Labour, etc.)

From the voting returns, we know the values of L, C, M and N, but not of p, q, r and s, which are the elements to be estimated.

The values of p, q, r and s can be estimated in the following way. Since all voters are either L or C, and are also either M or N, then

C = 1.0 − L; and
N = 1.0 − M.

We know that the proportion voting L in a constituency is the sum of the proportion of manual workers voting L plus that of non-manual workers voting L. Hence

L = (pM) + (rN); or
L = (pM) + r(1−M).

Simplifying, this becomes

L = (pM) + r − rM
 = r + (p − r)M

where (p−r) is the difference between the proportion of manual electors voting L and the proportion of non-manual electors voting L.

In a traditional regression context, using all constituencies as the observation units, it is possible to regress L on M, producing

L = a + bM

Since the constant value, a, is the value of L when M = 0, it is the proportion of electors voting L when there are no manual workers in the constituency; it is thus the value of r. If, on the other hand, the value of M were 1.0, this would be a constituency in which all of the voters are manual workers: thus a + b = p. If r and p are known, it is possible to calculate s and q too, since r + s = 1.0 and p + q = 1.0. Thus

r = a
p = a + b
s = 1 − a
q = 1 − (a + b)

so that if

L = 0.30 + 0.43M

then

r = 0.30 (30 per cent of non-manual electors vote Labour);
p = 0.73 (73 per cent of manual electors vote Labour);
s = 1 − 0.3 = 0.7 (70 per cent of non-manual electors vote
   Conservative);

and

q = 1 − 0.73 = 0.27 (27 per cent of manual electors vote
Conservative).

Thus from aggregate data it is possible to estimate the voting, by class, in each constituency.

The approach set out above produces the same values of p, q, r and s for every constituency, because it is an averaging procedure. However, the regression equation may be extended to

$$L = 0.30 + 0.43\ M + 0.10R$$

where

R is a regional dummy variable, set at 1 for constituencies in the north, and 0 for those in the south.

One can then estimate the values of
p_1, q_1, r_1, s_1, p_0, q_0, r_0, s_0

where

the subscript 1 represents a constituency in the north, and 0 a constituency in the south.

From the above equation

$p_1 = 0.83$ and $p_0 = 0.73$

indicating that the Labour Party obtained 10 percentage points more of the manual electors' votes in the north than in the south.

Inserting more independent variables, thereby increasing the goodness-of-fit of the equation and ensuring that it is properly specified, allows a range of spatial variations in voting by class to be estimated.

There are two basic problems with this approach, in the context of the hypotheses being tested here. The first is that although it is reasonably well suited to a situation with only two parties and either no abstentions or no social-spatial variations in the abstentions' rate, it is not particularly apt for analyses where there are three major parties, where the rate of abstentions is important and where the class breakdown involves more than two groups. Miller (1977, 1978) has introduced procedures based on the triangular graph that can incorporate a three-party situation (see also Upton, 1976 and Gudgin and Taylor, 1979), but their further extension is not straightforward.

Secondly, and more importantly, the procedure does not allow direct estimation of one of the major relationships hypothesised in the previous chapter. According to the first hypothesis set out there (p. 83), it is not that

$$L = a + bM,$$

but also that

$$p = a + bM, \text{ and}$$
$$q = c + dM$$

The regression approach does not disentangle the spatially-varying quantities of p and q suggested by this hypothesis.

To illustrate this point, standard ecological models have been estimated for the percentage of the votes won by each of the three main parties in 1983. Two sets of independent variables were used for each. The first was common to all three, and involved: eight dummy variables to represent eight of the nine English regions (the excluded region – in the constant term in the equations – was the Southwest); five dummy variables to represent five of the six constituency types (nonmetropolitan county constituencies were in the constant term – for the full list, see Table 5.1); four dummy variables representing constituencies in which an incumbent was standing; and a dummy variable set at 1 if the Alliance candidate was a member of the Liberal Party. The second set was particular to each model, and represented the 'best fit' set of ecological variables, identified after experimentation with several sets. To represent the basic social cleavage, the electorate was divided into socio-economic classes 1 and 2 (the white-collar classes in the census classification) and 3, 4 and 5 (the blue-collar classes), with the former used in the regression models for Conservative and Alliance voting and the latter for Labour voting. (Miller, 1977 and 1978, suggests – on quasi-theoretical grounds, plus empirical observation that it produces the best fit – that for both Conservative and Labour voting the percentage of the workforce classed as 'employers and managers' should be used, representing the 'controllers' in society. The experiments undertaken here showed that this variable was not superior as an explanatory variable, and interpretation was simpler using the chosen division.) To represent the consumption sector cleavage, the variable 'percentage of working-class residents in owner-occupied accommodation' was used; 'percentage employed in agriculture' and 'percentage employed in energy industries' were included in the Conservative and Labour models respectively to represent the relevant party's presumed strength in rural and mining areas. No other variables provided a significant contribution to a regression model estimating the Alliance vote.

The results of the regressions are shown in Table 5.1. For Conservative and Labour, the fit is very good (in each case, the three 'ecological' variables accounted for 77 per cent of the variation in regressions, excluding the general set of dummy variables). Even for the Alliance, the fit is substantial, and somewhat unexpected given the general interpretation of this group's 'universal' support (see p. 26); most of the variation was accounted for by the dummy variables, and the value of r^2 for an equation including only the variable SE12 was just 0.15.

Interpretation of these models confirms the findings of analyses of previous elections, at least for Conservative and Labour. All three parties increase their vote as the size of the social class from which they obtain most support increases. The Conservative Party gets more votes where the percentage of the working class living in owner-occupied houses is high; the Labour Party gets less. The Conservative Party increases its share of the vote as the percentage employed in agriculture increases; Labour's share increases with an increased percentage in energy industries. In addition, each party obtained an above-average share of the vote in constituencies where it fielded an incumbent, whereas within the Alliance, the Liberal candidates performed better than did those of the SDP.

Table 5.1 Standard Ecological Regressions of Voting: England, 1983 (n = 523)

Party	Conservative	Labour	Alliance
Constant	13.70	−14.10	24.37
Party Independents			
	SE12 0.56*	SE345 0.70*	SE12 0.12*
	W-Class O-O 0.33*	W-Class O-O − 0.31*	
	Agriculture 0.83*	Energy 0.40*	
Other Independents			
North	− 3.69*	7.54*	− 5.04*
Yorks/Humberside	− 3.25*	5.63*	− 4.19*
Northwest	− 2.47	9.02*	− 6.02*
West Midlands	− 1.97*	4.61*	− 5.78*
East Midlands	1.43	4.24*	− 6.27*
East Anglia	1.60	3.21*	− 4.08*
Greater London	− 1.19	13.40*	− 8.47*
Southeast	1.92*	4.36*	− 4.08*
Metropolitan Borough	− 1.69*	7.65*	− 2.23*
Inner London	− 3.73	3.98	0.58
Outer London	− 1.58	1.34	0.77
Metropolitan County	− 3.06*	5.97*	− 1.71
Nonmetropolitan Borough	0.25	4.46*	− 1.01
Conservative Incumbent	3.27*	− 2.86*	0.48
Labour Incumbent	1.21*	4.10*	− 2.46*
Liberal Incumbent	9.81*	−10.79*	19.89*
SDP Incumbent	− 2.14*	− 3.66*	6.40*
Liberal Candidate	− 0.29*	− 1.59*	1.35*
R² (adj.)	0.87	0.87	0.52

Note: *In this and all other tables in the book, an asterisk indicates that the regression coefficient is significantly different from zero at the 0.05 level.

There were also substantial inter-regional and urban:rural differences in vote-winning. Labour did significantly better in every region than in the Southwest, with its best average performances in the three northern regions and in Greater London; it also did significantly well in the metropolitan counties and in the borough constituencies in the nonmetropolitan areas. Complementing this, the Conservative Party did better in the Midlands and the Southeast than in the Southwest, but worse in the North; the Alliance performed best in the Southwest, and significantly below that level in every other region.

Using the regression coefficients from such equations, it should be possible to estimate the cell values of the following matrix, for any constituency

	Manual	Non-manual	
Labour	p	r	L
Conservative	q	s	C
Alliance	t	w	A
	M	N	

From the regression

$$C = a + bN$$

we can derive

$$a = q$$
$$s = a + b$$

So that 13.7 per cent of manual workers voted Conservative, compared to 69.7 per cent of non-manual workers.

Similarly, using the regression

$$L = a + bM$$

we could derive

$$a = r \quad \text{and} \quad p = a + b$$

but this would indicate a negative percentage of non-manual

voters supporting Labour: the estimated value of p is that 55.9 per cent of manual workers voted Labour.

Using the Alliance equation

$$A = a + bN$$

then

$$t = a \qquad w = a + b$$

suggesting that 24.37 per cent of the working class supported the Alliance, and 36.37 per cent of the middle class.

The full estimates are

$$
\begin{aligned}
p &= 55.9 & r &= ? \\
q &= 13.7 & s &= 69.7 \\
t &= 24.4 & w &= 36.4
\end{aligned}
$$

so that the manual voting is underestimated (6 per cent of the votes are unaccounted for) whereas the non-manual voting is overestimated (106.1 per cent, without the value of r)!

One could simplify the approach, reducing the matrix to a 2 × 2 format in the following way.

	Manual	Non-manual	
Conservative	q	s	C
Other	t	w	O
	M	N	

in which case we get

$$
\begin{aligned}
q &= 13.7 & s &= 69.7 \\
t &= 86.3 & w &= 30.3
\end{aligned}
$$

The full equation can then be used to estimate spatial variations in the values of q, s, t and w. Take the following two constituencies

	A		B	
WCOO	20	(6.6)	5	(1.65)

Ag.	8	(6.64)	1	(0.83)
East Anglia	1	(1.60)	0	(0)
North	0	(0)	1	(−3.69)
Metrop. Borough	0	(0)	1	(−1.69)
Con. Incumb.	1	(3.27)	0	(0)
Lab. Incumb.	0	(0)	1	(−1.21)

The basic estimates are altered by adding terms in brackets to q and s, and subtracting them from t and w. (The terms in brackets are the values of the independent variables multiplied by the relevant regression coefficients.) Thus s is equal to 87.81 in constituency A (an East Anglian agricultural constituency with a Conservative incumbent) and 65.59 in constituency B (a Northern, metropolitan borough seat with a Labour incumbent). As a consequence, w falls to 12.1. The values of q and t are altered in the same way: q becomes 31.81 and t 68.19.

The last paragraph indicates a substantial problem with this estimation procedure using regression models. The values of s and q are perfectly related, with the former always being 56 percentage points larger than the other. Regional variations in middle-class support for Conservative and working-class support for the same party are perfectly correlated, and no allowance for possible differences between the two can be made. Thus although the regression equations provide considerable insight into the geography of voting in England, the procedure used by Crewe and Payne:

(1) does not allow the percentage of a class voting for a party to vary with the relative importance of that class in a constituency, only with the relative importance of other variables;

(2) does not allow inter-class variations in differential support for a party, by constituency; and

(3) can produce estimates that are nonsensical (less than 0.0 or greater than 100.0).

Thus the procedure to be used here has been developed; it lacks all three of the above disadvantages.

Entropy-Maximising Estimation

Given a cross-classification, in which one knows the values of the row and column totals, but not of the individual cells, is it possible to estimate the latter? Entropy-maximising methods produce the maximum-likelihood estimate – the distribution that is most likely to occur given no other constraints than those imposed.

Take the following simple cross-classification, involving only six individual voters.

	M	N	
L	p	r	4
C	q	s	2
	3	3	

Four vote L, two C; there are three in each of M and N. What are the most likely values of p, q, r, s? Entropy-maximising methods identify the most-likely values as those which would occur most often, given the constraints. For example, assume that p = 3. If this is so, then r = 1, q = 0 and s = 2. Given six voters, (a, b, c, d, e, f,) there are 20 different sets of three voters (a,b,c; a,b,d; a,b,e, etc.) who could be in cell p. The other three must be distributed between r and s, with two in s and one in r. There are three ways in which this can occur. Thus in total there are 60 different ways in which the distribution p = 3, q = 0, r = 1, s = 2 could arise, each of which gives L = 4, C = 2, M = 3, N = 3. There are other possible distributions. If p = 2, for example, then q = 1, r = 2, s = 1; and so on.

Which is the most likely? Which is the distribution that occurs most often? (More formally, which distribution has the highest probability of occurring, given a random distribution of the voters among the four cells, subject to the constraints that the row and column totals must be those given in the table?) Perhaps not surprisingly, the answer is the distribution p = 2, q = 1, r = 2, s = 1; there are 180 different ways of distributing the six voters that produce this result.

The above example is a trivial one, with a small cross-classification and few voters. It can be extended, however. Instead of one constituency, let us have two, as follows.

	M_1	N_1	
L_1	p_1	r_1	5
C_1	q_1	s_1	2
	4	3	

	M_2	N_2	
L_2	p_2	r_2	3
C_2	q_2	s_2	5
	3	5	

The row and column sums are also given. With that information, which can be obtained from the census and election return data, the values of p_1, r_1, etc. could be estimated. They would be

2.86	2.14		1.13	1.87
		and		
1.14	0.86		1.87	3.13

(Note that these are not integer values. If integers were required the values would be 3, 1, 2, 1, and 1, 2, 2, 3 in the two matrices.)

In this exercise each constituency is being treated separately. But what if we also know the values of the internal cells? In this case, the matrix would be

$(p_1 + p_2)$	$(r_1 + r_2)$	$(L_1 + L_2)$
$(q_1 + q_2)$	$(s_1 + s_2)$	$(C_1 + C_2)$
$(M_1 + M_2)$	$(N_1 + N_2)$	

If $(p_1 + p_2) = P$, $(q_1 + q_2) = Q$ etc. then a further constraint could be introduced. As well as

$$p_1 + r_1 = L_1 \qquad p_2 + r_2 = L_2$$
$$p_1 + q_1 = M_1 \qquad p_2 + q_2 = M_2$$
$$q_1 + s_1 = C_1 \qquad q_2 + s_2 = C_2$$
$$r_1 + s_1 = N_1 \qquad r_2 + s_2 = N_2$$

in the estimation of the cell values, we would also have the constraints of

$$p_1 + p_2 = P \qquad q_1 + q_2 = Q$$
$$r_1 + r_2 = R \qquad s_1 + s_2 = S$$

so that if we know that

$$P = 6, Q = 2, R = 3, S = 4$$

then we would be estimating the values of p_1, p_2, etc. according to three sets of constraints (the rows, columns and faces of a data cube: Figure 5.1). The best estimates are then (in integers)

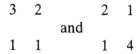

$$\begin{matrix} 3 & 2 \\ & & \text{and} \\ 1 & 1 \end{matrix} \qquad \begin{matrix} 2 & 1 \\ \\ 1 & 4 \end{matrix}$$

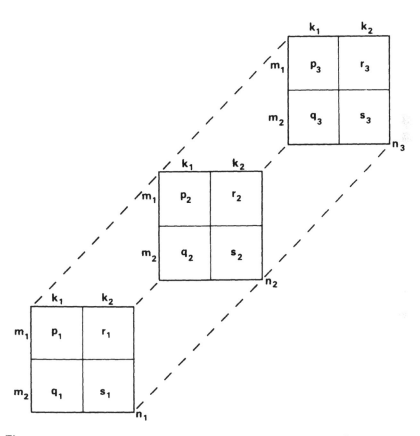

Figure 5.1 The Three-dimensional Data Cube mT^nk. In this example, there are three constituences ($n = 3$), two parties at the first election ($m = 2$), and two parties at the second election ($k = 2$). The known values (the constraints) are: the $_nS_m$ matrix, comprising the m row totals for each of the n constituencies; the $_nR_k$ matrix, comprising the k column totals for each of the n constituencies; and the $_mV_k$ matrix, comprising the sums of the p, q, r and s values across all n constituencies. The p_1, p_2, etc. values are to be estimated

(Note that these do not sum exactly to the values of P, Q, R, S, in part because of use of integers, and in part because of the approximation procedure used. The smaller the numbers, the greater the apparent 'error'. For further details, see Appendix 3.)

With entropy-maximising procedures, therefore, what we do is take the available information and use it to estimate unknown information, within the constraints that the values of the unknowns must sum to the knowns. As Wilson (1981, p. 59) displays it, the data can be recorded at three scales (Figure 5.2):

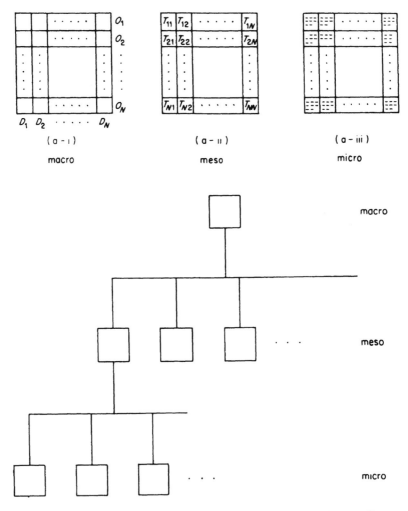

Figure 5.2 The Various States of a System, According to Wilson (1981, p. 59)

(1) At the *macro-scale*, we have the constraint values (the row and column totals shown by the dots in the left-hand, top matrix of Figure 5.2).

(2) At the *meso-scale*, we have the values in the cells of the matrix, which sum to the row and column totals (the dots in the central matrix).

(3) At the *micro-scale*, we have a distribution of the individual elements across the cells of the matrix; summing the entries provides the cell totals of the meso-scale.

There is, then, one macro-scale state, comprising the known information. A number of meso-scale states are associated with this, comprising the various ways in which the macro-state could be produced. And, finally, each meso-state has a series of micro-states associated with it, each of which produces the same distribution of cell values in the meso-state. Entropy-maximising methods assume that all micro-states are equally probable – i.e. there are no other constraints that might prevent some of them occurring. They identify the meso-state that has the most micro-states associated with it. This, given the assumption that the only constraints are those set out in the macro-state, is therefore the most likely meso-state. It is the most likely to occur and, as Wilson (1981, p. 59) expresses it

> Fortunately, the most probable state turns out to be very much the most probable.

Thus, if the constraints are properly identified, the values of the cells will be too. (A fuller proof is set out in Appendix 3.)

Using Entropy-Maximising Procedures in Voting Studies

Since the first presentation of these procedures to geographers in the late 1960s, they have been widely used to estimate certain flow matrices, using data about the origins and destinations and the costs of movement (see Wilson, 1970; Chilton and Poet, 1973; Senior, 1979). They have only recently been introduced to electoral studies (Johnston and Hay, 1982; though see a somewhat similar use in Boucharenc and Charlot, 1975, and also Wanat, 1982a, 1982b), however, and have since been used: to illustrate why the uniform swing observed in Britain cannot be the result of a spatially-invariant 'flow-of-the-vote' matrix (Johnston and Hay, 1982); to show that party loyalty was greatest, between the two 1974 general elections, in the marginal seats (Johnston and Hay, 1983); to illustrate sectional variations in the 'flow-of-the-vote' in New Zealand (Johnston, 1982b) and the United States (Johnston, 1984b); to show that there were spatial variations in the degree of 'ticket splitting' at the 1976 US elections (Johnston and Hay, 1984); to investigate the role of migration in voting change (Johnston, Hay and Taylor, 1982; Johnston, 1983a); and to test hypotheses such as the first two outlined above (p. 83), using 1960s and 1974 data (Johnston, 1982a, 1983b, 1983c). In the present context, they will be used to generate the dependent variables for the analyses outlined in the previous chapter.

Flow-of-the-Vote Matrices

The central role of the flow-of-the-vote matrix (or voter transition matrix as it is sometimes called) in electoral change has led several analysts to seek methods of identifying both the national matrix and constituency matrices, using aggregate data. Miller (1972), for example, used regression models to predict loyalty and switching votes, using both the voting pattern at the first of the pair of elections and constituency environment data to predict the result at the second election (see also Hawkes, 1969). As with the regression approach used by Crewe and Payne (1976), however, this procedure does not allow for spatially-varying values of the matrix elements, and so cannot be used to test the hypotheses outlined here.

An alternative procedure uses mathematical programming procedures (Irwin and Meeter, 1969). This has been applied to British data by McCarthy and Ryan (1977; see also McCarthy and Ryan, 1976). A matrix for a set of constituencies (either the country as a whole or some subset of it) is constructed from the row and column sums of each constituency matrix, on the assumption that the same matrix applies in all places. The quadratic programming procedure used identifies the aggregate matrix that fits all of the individual matrices. Its results, unlike those obtained via the regression approach, must all be non-negative, but, as Upton (1978a) has shown, compared to the evidence from survey data it very much overemphasises party loyalty and underestimates the amount of inter-party movement. Separate analyses for groups of constituencies allow for testing differences between them in the relative size of the cell entries, but it is not possible to test whether there are differences between individual constituencies (see also McLean, 1973).

The hypotheses presented here are that there are substantial inter-constituency variations in the elements of the flow-of-the-vote matrix. The degree of party loyalty, for example, should be related to both the strength of the party at the first election and the marginality of the constituency. Rather than estimate flow-of-the-vote matrices for groups of constituencies (safe Labour; marginal; etc.) using either of the above methods, the entropy-maximising procedure was adopted because it provides an estimate of the flow in each constituency, consistent with both the two election results and the national – survey-estimated – flow-of-the-vote matrix.

The basis of the procedure is easily stated. Given a set of n constituencies and two elections, with m parties contesting the first election and k parties contesting the second (note that m need not be the same as k, and that abstention counts as a separate party), one needs:

(1) the election result in each of the n constituencies, at each of the two elections; and
(2) an estimate, obtained from a national sample survey, of the flow-of-the-vote matrix for all constituencies.

One can then estimate the values of the flow-of-the-vote matrix in each constituency.

The givens are

$_n S_m$ – a matrix of n rows (constituencies) and m columns (parties), showing the result of the first election in each constituency;

$_n R_k$ – a matrix of n rows (constituencies) and k columns (parties), showing the result of the second election in each constituency; and

$_m V_k$ – a matrix of m rows (parties) and k columns (parties), showing the flow-of-the-vote between the two elections, across all constituencies

The three matrices provide the faces of a data cube (Figure 5.1)

$_m T^n_k$ – a three-dimensional matrix, comprising m rows (parties), k columns (parties) and n faces (constituences).

Thus

$$\sum_{i=1}^{n} \sum_{j=1}^{m} S_{ij} = \sum_{i=1}^{n} \sum_{o=1}^{k} R_{ij} = \sum_{j=1}^{m} \sum_{o=1}^{k} V_{ij} = \sum_{i=1}^{n} \sum_{j=1}^{m} \sum_{o=1}^{k} T_{ijo}$$

The values of T_{ijo} (the number of voters who voted for party j at election 1 and party o at election 2, in constituency i) are unknown. Maximum-likelihood estimates are produced, using the entropy-maximising procedure, whereby the values of T_{ijo} are constrained to fit the sums in the matrices S, R and V.

$$\sum_{o=1}^{k} T_{ijo} = S_{ij} \quad , \text{ for all i and j}$$

$$\sum_{j=1}^{m} T_{ijo} = R_{io} \quad , \text{ for all i and o}$$

$$\sum_{i=1}^{n} T_{ijo} = V_{jo} \quad , \text{ for all j and o}$$

The first of these constraints requires that the sum of the movements from a party, in a constituency, to all others equals the total vote received by that party at the first election there. The second requires that the sum of the movements to a party in a constituency, from all others, equals the total vote received by that party at the second election there. And the third requires that the sum of the movements from one party to another, across all constituencies, equals the total volume of movement between the two.

The estimated constituency flow-of-the-vote matrices are thus the best estimates of the gross pattern of shifts, given both the two election results and the nationally-estimated flow matrix. The procedure used to produce the estimates is a simple one, involving an iterative process. (Full details are given Appendix 3.) Complete fits to the constraints are rarely achieved (in part because of 'error' in the data comprising the constraints, notably in the national flow-of-the-vote matrix). The criterion applied in all of the estimations undertaken here is that the best-fit has the smallest overall average deviation from the constraints, with no average deviation exceeding 5 per cent. In the iterative procedure, at each step one of the constraints is exactly met. Thus, for example, it may be that the third of the constraints is exactly met (i.e. the individual flows all sum to the national flows). On the others, the average deviation is calculated, viz.

$$D_S = [\sum_{i=1}^{n} \sum_{j=1}^{m} ((\hat{S}_{ij} - S_{ij})/S_{ij})]/(nm)$$

where

> S_{ij} is the number of votes received by party j in constituency i, at election 1;
>
> $\hat{S}_{ij}$ is the estimated number of votes received by party j in constituency i, at election 1, so that

$$\sum_{0=1}^{k} \hat{T}_{ijo} = \hat{S}_{ij}$$

where

> $\hat{T}_{ijo}$ is the estimated number of voters in constituency i for party j at election 1 and party o at election 2

and

> D_S is the average deviation for matrix S.

A similar value of D_r can be calculated. At the next step of the procedure, either D_r or D_S will be 0.0, and a value of D_v will be calculated as

$$D_V = [\sum_{j=1}^{m} \sum_{o=1}^{k} (\hat{V}_{jo} - V_{jo})/V_{jo})]/(mk)$$

where

$$\sum_{i=1}^{n} \hat{T}_{ijo} = \hat{V}_{jo}$$

The procedure continues until D_S, D_r and D_V are less than 0.05 and the two non-zero elements are as small as possible. (The procedure usually enters a loop after only a few iterations, repeating previously obtained 'solutions'.)

Clearly, this procedure is dependent upon the validity of the surveyed national matrix for the accuracy of its constituency elements. As suggested earlier (Chapter 2), such matrices when applied to two election results tend to underestimate the abstentions at the second election and to overestimate voting. (This signifies that there is a tendency for people to claim that they voted rather than admit they abstained.) To counter this, it is necessary to 'smooth' the matrix, using the procedure outlined in Appendix 3. Without this, the requirements that

$$\sum_{o=1}^{k} V_{jo} = \sum_{i=1}^{n} S_{ij}$$

and

$$\sum_{j=1}^{m} V_{jo} = \sum_{i=1}^{n} R_{io}$$

would not be met: the total number of votes for party j at election 1, according to the flow-of-the-vote matrix, would not equal the total number according to the national election result; and the total number of votes for party o at election 2 similarly would not be the same in both flow-of-the-vote matrix and election result.

With a 'balanced' national flow-of-the-vote matrix – i.e. one in which application of the percentages in the rows to the row totals produces the expected column totals – one can then proceed to estimate the T_{ijo} values. (The procedures are outlined in Appendix 3.) The output is a set of flows which produces a perfect fit to one of the constraints and is the best fit to the others. The following example illustrates this. There are two constituencies and three parties. The matrices are

S	100	150	R	120	170	V	200	40	10
	120	70		90	60		70	100	20
	60	80		70	70		20	10	110

Table 5.2 shows the results of the iterations seeking the best estimates, within the error limit allowed here (an average standardised deviation of 0.05). The first iteration, as always, produces poor fits. The second, using the V matrix totals as the denominators, produces perfect fits on the row totals (the S matrix), but an average standardised deviation of 0.09 (i.e. 9 per cent) on the column (the R matrix). The third iteration, using the column sums as the denominators, produces an acceptable fit on both the row totals (S matrix) and the V matrix. In the latter, the cell values expressed as percentages of the row totals are

80.0	16.0	4.0
36.8	52.6	10.6
14.3	7.1	78.6

The two estimated matrices are

i=1			i=2		
77.0	18.3	4.7	82.7	13.8	3.5
32.7	55.9	11.3	41.1	48.9	10.0
12.1	7.2	80.7	16.4	6.8	76.8

These differ one from the other, and from the 'national' matrix with, for example, greater loyalty to the first party in constituency 2 than there was nationally, but lesser loyalty to the other two parties. (With only two constituencies, of course, the 'national' figure is bound to be roughly the average of the two.)

Using the matrix in Table 2.4, estimates of the flow-of-the-vote in each of the 523 English constituencies have been

Table 5.2 A Hypothetical Iteration*

First Iteration

i = 1

c1	c2	(row)
33.3	33.3	100
40	40	120
20	20	60
93.3	93.3	93.3
(120)	(90)	(70)

i = 2

c1	c2	c3
50	50	50
23.3	23.3	23.3
26.7	26.7	26.7
100	100	100
(170)	(60)	(70)

∨ (aggregate)

83.3 (200)	83.3 (40)	83.3 (10)
63.3 (70)	63.3 (100)	63.3 (20)
46.7 (20)	46.7 (10)	46.7 (110)

average standardised deviation: columns .35   V 1.91

Second Iteration

i = 1

c1	c2	c3	(row)
80	4	16	100 (100)
44.2	12.6	63.2	120 (120)
8.6	47.1	4.3	60 (60)
132.8	63.7	83.5	
(120)	(70)	(90)	

i = 2

c1	c2	c3	(row)
120	24	6	150 (150)
25.8	36.8	7.4	70 (70)
11.4	5.7	62.9	80 (80)
157.2	66.5	76.3	
(170)	(60)	(70)	

average standardised deviation: rows OO columns 0.09

Third Iteration

i = 1

c1	c2	c3	(row)
72.3	17.2	4.4	93.9 (100)
39.9	68.2	13.8	121.9 (120)
7.8	4.6	51.8	64.2 (60)
120	90	70	

i = 2

c1	c2	c3	(row)
129.8	21.7	5.5	157 (150)
27.9	33.2	6.8	67.9 (70)
12.3	5.1	57.7	75.1 (80)
170	60	70	

∨ (aggregate)

202.1 (200)	38.9 (40)	9.9 (10)
67.8 (70)	101.4 (100)	20.6 (20)
20.1 (20)	9.7 (10)	109.5 (110)

average standardised deviation: rows .047   V 0.018

Note: *The values in brackets are the actual (constraint) values.

produced, to provide dependent variables for analyses of the geography of electoral change between 1979 and 1983.

Variations in the Basic Cleavages

Matrices cross-classifying social characteristics by voting can be analysed in exactly the same way. Given, say, a national matrix, derived from a survey, that cross-classifies occupational status against vote in 1983 (as in Table 2.1), plus data on the occupational composition and voting in each constituency, then the entropy-maximising procedure can be used to predict the number of votes for each party among members of each occupational class, in each constituency. Data for the occupational classification are not available for 1983, so the 1981 census data are used, assuming that there was no shift over the intervening two years. (The percentage distribution in 1981 was applied to the 1983 electorate to provide the assumed numbers in each class in 1983.)

This procedure has been used to estimate voting by class and housing tenure, in constituencies, at earlier elections (Johnston, 1982a, 1983b, 1983c). It has also been applied to a Western Australian data set, with which sensitivity testing was possible because the survey data provided the matrix for each of the ten constituencies in which electors were sampled. The results showed a very close fit (Johnston, Hay and Rumley, 1983; see also Rumley, 1981), providing some validation for the procedure.

With these estimates, it is possible to test the hypotheses relating the proportion of voters in a given category supporting a particular party to the size of that party's 'core group' in the relevant constituency (as in hypotheses 1 and 2: p. 83). This allows preciser evaluation of the local effects hypotheses than has been possible before. (There are potential problems of some circular reasoning in this, because the estimates are derived, in part, as a function of the population which later forms the major independent variable in the analysis. The nature of these problems, and of the relevance of the procedure, is discussed in Johnston, Hay and Rumley, 1984. Appendix 3 summarises the argument.)

The procedure can be used to evaluate multiple classifications. For example, it may be desirable to estimate voting behaviour separately for each occupational class according to housing tenure. Thus the national matrix, V, may have rows and columns as follows:

Occupational Class	Housing Tenure	C	L	Vote All	Ab
White-Collar	Owner-Occupier Other				
Blue-Collar	Owner-Occupier Other				

A similar matrix may be possible for each constituency, too, if the cross-classification of occupational class against housing tenure is available (it is in the 1981 Census). If it were not, then the constraints would have to be organised differently. The matrix could be arranged as follows:

Occupational Class Housing Tenure		Middle OO	Other	Working OO	Other
Constituency 1	C	a	b	c	d
	L	e	f	g	h
	A	i	j	k	l
	Ab	m	n	o	p
		1	2	3	4

The row columns are known, so that in the entropy-maximising procedure (a + b + c + d) would have to equal the Conservative vote in the constituency, (e + f + g + h) the Labour vote, and so on. The four column totals are not known. However, the total number of owner-occupiers is known, as is the total number of white-collar voters, etc. Thus the constraints are

$$(a + c + e + g + i + k + m + o) = (1 + 3)$$
$$(b + d + f + h + j + l + n + p) = (2 + 4)$$
$$(a + b + e + f + i + j + m + n) = (1 + 2)$$
$$(c + d + g + h + k + l + o + p) = (3 + 4)$$

so that after each iteration the estimates are summed over three of the four constraints (the national matrix, V; the vote matrix; the occupational class matrix; and the housing tenure matrix). This is the same as the treatment of migration (Johnston, Hay and Taylor, 1982).

In Summary

The hypotheses to be tested in this book require data on: (1) voting by social characteristics in each constituency; and (2) the 'flow-of-the-vote' in each constituency. Such data are not available. Previous attempts to estimate them have been limited by the procedures employed. Here, entropy-maximising procedures are used to produce the maximum-likelihood (most probable) estimates, which are in the exact format needed for the tests.

6 THE GEOGRAPHY OF VOTING, BY CLASS, 1983

The brief analyses in Chapters 2 and 3 indicated clear geographical variations in the propensity of the members of different occupational, social class and tenure categories to vote for the three parties in 1983. This chapter presents estimates of those variations, using the entropy-maximising procedure outlined in the previous chapter, and in a series of regression analyses investigates their geography.

Social Class and the Geography of Voting

The national matrix cross-classifying social class by voting in 1983 is given in Table 2.3. Grossed up to the national electorate, after smoothing to remove the underestimated non-voting, this matrix provides one of the constraints for the entropy-maximising estimation procedure; the other two are a matrix of number of electors in each social class by constituency and a matrix containing the 1983 election result. Estimates of voting by class in each of the 523 constituencies, with average deviations of less than 5 per cent, were obtained after only three iterations. For the analyses here, the number of votes was expressed as a percentage of the relevant row sum in each constituency (e.g. percentage of members of social classes I and II voting Conservative).

The frequency distributions for the 16 cell percentages, over all 523 constituencies, are given in Figures 6.1-6.4, with the summary statistics in Table 6.1. A brief glance at these indicates how substantial the spatial variations in voting behaviour must have been. Only two of the 16 coefficients of variation are less than 0.20, and in every one the range of values is at least 30 percentage points.

Comparing the social class groupings, there are few very substantial differences in the range and degree of variability; the main exception is with social classes IV and V, which were more variable, in relative terms, than the others in their propensity to

Table 6.1 Estimated Voting by Social Class: Summary Statistics

Social Class	1983 Vote	National Figure	Constituency Mean	Standard Deviation	Coefficient of Variation	Minimum	Maximum
I and II	Conservative	51.7	47.8	9.9	0.21	14	63
	Labour	10.2	11.6	7.0	0.60	1	32
	Alliance	20.2	20.0	4.8	0.24	7	44
	Did Not Vote	17.9	20.5	5.5	0.27	12	48
IIIN	Conservative	45.1	43.3	10.2	0.24	12	60
	Labour	15.7	15.5	8.7	0.56	1	39
	Alliance	17.9	18.1	4.8	0.27	6	42
	Did Not Vote	21.3	23.1	5.3	0.23	15	50
IIIM	Conservative	27.6	28.6	8.4	0.29	6	43
	Labour	24.2	20.2	10.2	0.50	2	44
	Alliance	18.2	19.2	5.8	0.30	5	46
	Did Not Vote	30.0	32.1	5.3	0.17	23	58
IV and V	Conservative	19.3	20.7	7.0	0.34	4	33
	Labour	31.3	26.3	12.4	0.47	3	52
	Alliance	17.7	19.1	6.5	0.34	5	48
	Did Not Vote	31.7	33.9	5.0	0.15	22	58

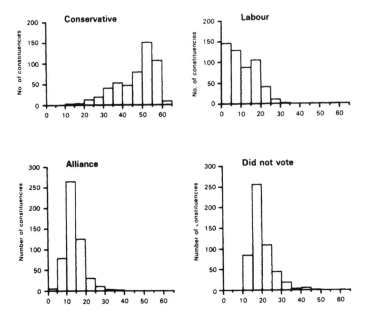

Figure 6.1 Frequency Distributions of Estimated Voting by Members of Social Classes I and II

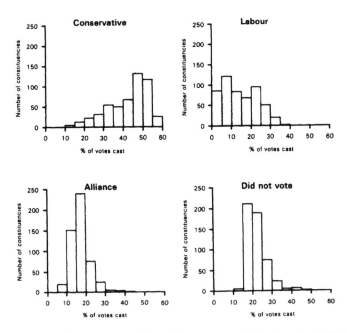

Figure 6.2 Frequency Distributions of Estimated Voting by Members of Social Class IIIN

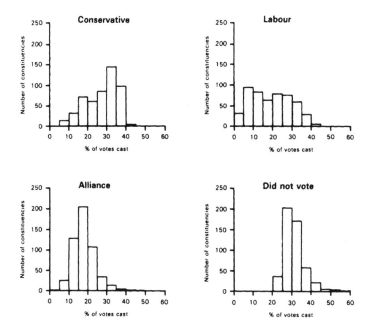

Figure 6.3 Frequency Distributions of Estimated Voting by Members of Social Class IIIM

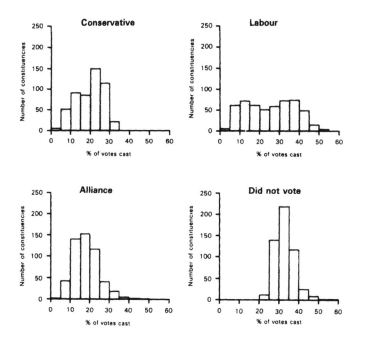

Figure 6.4 Frequency Distributions of Estimated Voting by Members of Social Classes IV and V

vote either Conservative or Alliance, but not Labour. There are substantial differences in the shapes of some of the distributions, however. For example, that for voting Labour among social classes I and II is strongly positively skewed, whereas for social classes IIIN and IIIM the same distribution is almost rectangular.

Much more apparent than differences between the classes are differences between the parties. The coefficients of variation for voting Labour are more than twice those for the other parties in the two white-collar classes, and they are substantially larger in the two blue-collar classes as well. Compared to the other parties, the likelihood of a member of any social class voting Labour varies very substantially across the 523 constituencies.

Regression Models of the Pattern of Class Voting

The preceding section has indicated very significant geographical variability in the pattern of voting by social class, according to the estimates produced by the entropy-maximising procedure. The present section outlines a regression model for identifying the parameters of that variability, based on the hypotheses presented in Chapter 4. (The use of regression models in this context is discussed in Appendix 4.)

According to most analyses of geographical variations in voting in England, the propensity for the members of any social class to vote for a particular party should be a function of the size of that party's core group of supporters in the constituency. Thus, for example, if the Conservative Party's core group of supporters (i.e. those most likely to vote for it) are those in managerial occupations, then the greater the percentage of a constituency's population in managerial households the greater the percentage voting Conservative, in every social class.

Which, then, are the core classes? Miller's (1977) identification of these was largely empirical, to which a theoretical interpretation was given. In general, the Conservative Party gets more support from white-collar workers and from owner-occupiers (see Table 2.3) whereas Labour gets most from blue-collar workers and from council tenants. For the present purposes, a variety of class, tenure and class plus tenure variables were regressed against the estimated voting percentages to identify the best fits, for which there were sensible interpretations. These led to the following selection:

Conservative – the percentage of the electorate living in social class I and II, owner-occupied households;

Labour – the percentage of the electorate living in social class IIIM, IV and V households; and

Alliance – the percentage of the electorate living in social class II households.

For the regression model, therefore, it is hypothesised that: (1) the larger the percentage in middle-class 0-0 households (social class I and II, owner-occupied), the greater the Conservative vote in 1983, in all classes; (2) the larger the percentage in working-class households (social class IIIM, IV, V), the greater the Labour vote, in all classes; and (3) the larger the percentage in social class II households, the greater the Alliance vote, in all classes. These hypotheses are consistent with the contagious and sectional effects.

Analyses have shown that, holding constant the class composition of the electorate, other aspects of a constituency's socioeconomic structure influence the pattern of party voting. Five variables are introduced into the present model to represent these potential influences.

1. The Percentage of the Workforce Employed in Agriculture. It is generally assumed that agricultural workers are more likely to vote Conservative, and that there is a general pro-Conservative ethos in rural areas; these reflect a variety of influences, such as the long tradition of protectionist policies by the Conservative Party, which favours agricultural interests because it shields them from cheap imported competition, and the deferential character of rural workers (Newby, 1977). Thus it is hypothesised that the greater the percentage of an area's workforce employed in agriculture the greater the support for the Conservative Party, in all classes, and the smaller the support for Labour and the Alliance.

2. The Percentage of the Workforce Employed in Energy Industries. The coalfields have been the major area of support for the Labour Party for more than 50 years, based, it is assumed, not only on the strongly working-class character of those areas but also on the closely-knit nature of the mining communities which socialises most members into a pro-Labour stance (Crewe, 1973). The census data do not give separate figures for the percentage employed in coal-mining, so the percentage employed in energy industries (which includes the power industry with mining) is used as a valid surrogate; it is hypothesised that the larger this percentage the greater the support for Labour, from all social classes.

3. The Percentage of the Workforce Employed in Manufacturing. In much of England, most support for Labour has come from industrial workers, especially those employed in large factories and in the more unionised occupations. Thus, holding social class constant, it is hypothesised that the larger the percentage of the workforce employed in manufacturing, the greater the support for Labour, in all classes.

4. The Percentage of the Workforce Unemployed. Governments are generally blamed when a country is suffering economic depression, and if the depth of that depression is spatially variable then the extent of discontent with government should be too. (Many opinion polls in recent years have shown that unemployment has been a major concern of the British electorate: Budge and Fairlie, 1983, illustrate this in their work on predicting election results. See also Johnston, 1983d.) A reasonable indicator of the geography of the depression is given by the percentage of the workforce that is unemployed (although the census data refer to 1981 and not 1983). Since the Conservative Party was in power prior to the election, it is hypothesised that the higher the level of unemployment in a constituency, the smaller the percentage voting Conservative, in all social classes.

5. The Percentage of Working-class Households Living in Owner-occupied Dwellings. The embourgeoisement thesis (p. 17) is that home ownership encourages Conservative voting, especially among the working classes; this has been capitalised on by the Conservative Party in recent years with its policies promoting the sale of council houses. Thus it is hypothesised that the larger the percentage of a constituency's working-class households in owner-occupied dwellings, the greater the Conservative vote among the working class and the smaller the Labour vote. Because the variable refers to the situation in 1981, it cannot capture the effects of recent policies on council-house sales very easily. Hence (as suggested in Appendix 2) its main focus is on longer-term variations in working-class owner-occupancy.

In addition to these five variables which are constituency-specific, further geographical effects are suggested, following the identification of regional and urban:rural cleavages by Curtice and Steed (1982). For the regional differences, England was divided into the nine OPCS regions:

North	West Midlands	Greater London
Yorkshire and Humberside	East Midlands	Southeast
Northwest	East Anglia	Southwest

The first eight of these were represented by dummy variables in the regression models; the ninth – Southwest – was incorporated in the constant term. For the urban:rural cleavage, constituencies were split into six groups representing the metropolitan: nonmetropolitan division of counties; the borough and county classification of constituencies by the Boundary Commission (reflecting population density); and a division of Greater London into Inner (the ILEA boroughs) and Outer to reflect inner city:suburban differences (as in the borough:county division within metropolitan counties). (The urban:rural division is measured by Steed and Curtice, 1983, as the population density; this is a finer index than used here, but employment in agriculture is also used here.) This gives:

Metropolitan Borough	Metropolitan County
Inner London	Nonmetropolitan Borough
Outer London	Nonmetropolitan County

of which the first five were represented by dummy variables and the sixth – Nonmetropolitan County – was incorporated into the constant term. (See Appendix 2 for maps of the classifications.)

With these dummy variables it was hypothesised that:

(1) Compared to the Southwest, voting for Labour would be significantly greater in every region, among all classes, with the greatest difference being in the northern regions. Conversely, voting for Conservative and the Alliance would be significantly smaller.

(2) Compared to the Nonmetropolitan County constituencies, voting for Labour would be significantly greater in all five types, among all classes, with the greatest difference in the Metropolitan Borough and Inner London constituencies. Conversely,

voting for Conservative and the Alliance would be significantly smaller there. (On the possibility of collinearity effects, see Appendix 4.)

Finally, six variables were introduced relating to the constituency contest. The first four cover incumbency. Electoral studies have shown that in the American political system, where party is much less important than it is in England, there is a very strong incumbency effect. British evidence on this is far from clear. However, it is hypothesised here that there is an incumbency effect, and that, *ceteris paribus*, incumbents win more votes from all social classes than do non-incumbents. To test this, four dummy variables were incorporated in the regression model, representing Conservative, Labour, Liberal and Social Democrat incumbent candidates; because the 1983 election was being fought in new constituencies, an incumbent was defined as a candidate who had previously held a seat covering either part of the constituency now being contested or an adjacent one.

The 1983 election introduced a new situation for the vast majority of British voters: two parties – Liberal and Social Democrat – were contesting the election as an Alliance, and had agreed which party would provide the candidate for each constituency. Their electoral base was the Liberal vote in 1979, and SDP candidates had to win over the former Liberal supporters – for which they had the backing of the Liberal Party. In addition, the two Alliance Parties had to win converts from the other two parties, and from those who did not vote in 1979. A number of the SDP candidates – including its four leaders ('The Gang of Four' – Roy Jenkins, David Owen, Bill Rodgers and Shirley Williams) and all but one of its incumbent MPs – had defected from the Labour Party; they had to establish their new credentials and withstand charges of 'traitors' and 'unethical' (all but one of the defectors declined to resign and fight a by-election) behaviour. For these reasons, some commentators suggested that Liberal candidates for the Alliance would win more votes than would SDP candidates; this hypothesis was tested by the use of a dummy variable, coded 1 for the constituencies fought for the Alliance by the Liberal Party, and 0 otherwise.

Although in general local issues are thought to influence electoral behaviour in England very little, one aspect of the constituency that does apparently have some effect is the marginality of the seat. The closer the contest, the less likely that voters will desert their 'traditional' party, especially to cast a 'protest vote' – either for a third party or to abstain. Thus it was hypothesised that the more marginal a seat the greater the adherence to class norms in the voting. Marginality was measured as the vote difference between the Conservative and Labour Parties in the estimated 1979 results, expressed as a percentage of the electorate. (This is not an ideal measure, because in about 60 constituencies the estimated 1979 votes put Conservative and Liberal in the top two places. It proved impossible to devise a single measure of marginality to cater for this in the context of later studies of the flow-of-the-vote, however, so a measure using the 'traditional' two main parties was included, on the implicit assumption that, as national issues are the most important to the electorate, it was the relative positions of these two which influenced votes.)

In total, therefore, each regression model contained 25 independent variables. After testing for the absence of any substantial multicollinearity these were entered *en bloc*, and all of the regression coefficients are reported here. The R^2 measure of goodness-of-fit was adjusted to take account of the loss of degrees of freedom involved in the relatively large number of variables (Nie *et al.*, 1975; Draper and Smith, 1966); the statistical significance of the coefficients (Appendix 4) is indicated in the tables by starring of those significant at the 0.05 level.

The Results

The twelve regression equations are reported in Tables 6.2-6.4. Overall, they provide substantial evidence in support of most of the hypotheses being tested here. The average R^2 value is 0.795, indicating that almost 80 per cent of the spatial variations in voting by social class can be accounted for in terms of the independent variables established here. For both Conservative and Labour voting patterns the average R^2 is 0.9; voting for the Alliance was much less predictable.

Looking first at the results for the *Conservative Party* (Table 6.2), the first block of coefficients shows the expected pattern with regard to the assumed effect of the 'core' voters. The greater the proportion of middle-class owner-occupiers in the constitu-

Table 6.2 Regression Analyses: Percentage in each Social Class Voting Conservative

Social Class	I and II	IIIN	IIIM	IV and V
Constant	38.16	34.07	20.71	13.98
Middle Class O-O	1.12*	1.12*	0.88*	0.71*
Agriculture	0.59*	0.61*	0.51*	0.43*
Energy	−0.27*	−0.30*	−0.24*	−0.21*
Manufacturing	0.04	0.04	0.05	0.04
Unemployment	−0.39*	−0.44*	−0.38*	−0.31*
Working Class O-O	0.02	0.01	−0.01	−0.01
Marginality	−0.14*	−0.11*	−0.04*	−0.01
North	−1.55	−2.11*	−2.07*	−2.01*
Yorks/Humberside	−2.10*	−2.48*	−2.23*	−1.97*
Northwest	−0.24	−0.82	−0.89	−0.99*
West Midlands	2.52*	1.98*	1.13	0.67
East Midlands	2.46*	1.92*	1.12	0.70
East Anglia	0.78	0.30	0.10	−0.08
Greater London	−0.58	−1.37	−1.39	−1.21
Southeast	1.35*	0.97	0.57	0.40
Metropolitan Borough	−2.58*	−2.73*	−2.37*	−1.91*
Inner London	−4.62	−4.49	−3.53*	−2.78*
Outer London	−2.84	−2.74	−2.35	−1.95
Metropolitan County	−2.07*	−2.22*	−2.05*	−1.74
Nonmetropolitan Borough	−1.08*	−1.19*	−1.18*	−0.99*
Conservative Incumbent	2.56*	2.52*	2.01*	1.57*
Labour Incumbent	−0.73	−1.15*	−1.16*	−1.06*
Liberal Incumbent	−9.40*	−7.82*	−5.39*	−3.83*
SDP Incumbent	−2.69*	−2.11*	−1.33*	−0.84*
Liberal Candidate	−0.28	−0.15	−0.04	0.02
R^2 (adj.)	0.88	0.90	0.91	0.92

ency electorate, the greater the propensity of members of all class groups to vote Conservative; among the middle classes (I, II and IIIN) the coefficients greater than 1.0 indicate that members are readily socialised into the local context, whereas for the working class (IIIM, IV and V) the coefficients less than 1.0 suggest that the willingness to cross the traditional cleavage lines is less than the willingness among the middle class to express class solidarity when the bulk of the local environment promotes the Conservative Party ideology (exactly the same happens, in the opposite direction, with voting Labour – Table 6.3 – though to a lesser degree). In general, then, the more pro-Conservative the constituency the more prepared members of all classes were to vote Conservative in 1983 with not surprisingly the middle class more prepared than the working class.

Of the five variables representing various aspects of constituency socio-economic structure, three have significant regression coefficients – in all four equations. All of the signs for these three are as predicted. Thus the more people employed in agriculture in a constituency, the greater the Conservative vote, especially among the middle class, and the more there are employed in energy industries the smaller the Conservative vote, especially among the working class. Thirdly, the higher the level of unemployment the smaller the Conservative vote – slightly more so among the middle than the working class. There were no significant relationships involving either employment in manufacturing or, more surprisingly, working-class owner-occupancy.

The eight regional variables were mostly insignificant in the equations. Only for the Yorkshire and Humberside region is there consistency across all four; compared to voters in the Southwest, about 2 per cent less voted Conservative in all social classes (as was the case in three of the four classes in the Northern region). The other salient block of regression coefficients refers to the West and East Midlands, where middle-class voters supported the Conservative candidates in significantly greater proportions than did their counterparts elsewhere in the country.

In contrast to the absence of marked regional cleavages in class support for the Conservative Party, the regression coefficients suggest a major urban:rural cleavage. Compared to its performance in the nonmetropolitan county constituencies, the party did poorly in the metropolitan counties and also in the nonmetropolitan borough constituencies; in general, residents of urban England were 1 to 2 per cent less likely to support the Conservative Party than were their contemporaries in the shires. The exception to this comes with Greater London. The regional variable for London was insignificant in all four, and although all of the coefficients for Inner and Outer London were both substantial and negative, only two, referring to working-class voters in Inner London, were statistically significant. The reason for this (matched in the other tables) is the relative success of the Alliance in certain parts only of both Inner and Outer London, as discussed below; each party achieved success in parts of both Inner and Outer London, but also suffered some major setbacks.

Turning to the variables relating to the local contest, the results for marginality are as expected. According to the hypothesis, the safer the seat the greater the likelihood of middle-class voters deserting the Conservative Party. The negative regression coefficients support this; the two coefficients for middle-class voters are the largest, as expected, and one of the two for the working class is statistically insignificant. The more marginal the seat, therefore, the greater the loyalty of the middle-class voters to their 'natural party'.

Regarding incumbency, ten of the twelve regression coefficients are statistically significant and all have the anticipated sign. Members of all classes were more likely to support a Conservative incumbent than either a 'freshman' or one defeated before, and if an incumbent of another party was standing they were less likely to vote Conservative. Regarding the latter finding, Liberal incumbents were especially successful in winning what otherwise might have been Conservative votes. (This is true in absolute terms. However, the 9.4 per cent less votes given by classes I and II to Conservative where a Liberal candidate was standing is 25 per cent of the constant term – i.e. the Conservative vote when no incumbents are standing – whereas for classes IV and V it is 27 per cent. Relative to what might be expected, the Liberal incumbents won support almost equally from all four classes.) With the Alliance, there was no difference in the Conservative vote if a Liberal Party candidate was standing.

Some of the results shown for the *Labour Party* (Table 6.3) not surprisingly complement those just described. Thus the more working class the constituency, the greater the propensity to vote Labour among all classes; in relative terms, however, a strong pro-Labour presence is more likely to win votes among the working class. Where the working class comprises 30 per cent of the electorate, Labour wins on average less than 3 per cent of the votes of classes I and II compared to 16 per cent of those in classes IV and V; for constituencies where the working class makes up 60 per cent of the electorate, the respective percentages are 8 and 32.

Table 6.3 Regression Analyses: Percentage in Each Social Class Voting Labour

Social Classes	I and II	IIIN	IIIM	IV and V
Constant	−2.72	−1.39	3.49	8.35
Working Class	0.18*	0.22*	0.23*	0.26*
Agriculture	−0.23*	−0.31*	−0.37*	−0.45*
Energy	0.39*	0.48*	0.56*	0.64*
Manufacturing	0.08*	0.11*	0.13*	0.16*
Unemployment	0.35*	0.40*	0.41*	0.44*
Working Class O-O	−0.13*	−0.15*	−0.16*	−0.17*
Marginality	−0.06*	−0.09*	−0.18*	−0.25*
North	2.62*	3.59*	4.70*	5.98*
Yorks/Humberside	2.17*	2.81*	3.30*	4.13*
Northwest	3.33*	4.25*	5.26*	6.45*
West Midlands	0.20	0.42	0.91	1.48
East Midlands	0.23	0.42	0.78	1.27
East Anglia	1.34*	1.69*	2.13*	2.81*
Greater London	4.07*	4.98*	5.42*	6.24*
Southeast	0.82*	0.97	1.07	1.29
Metropolitan Borough	1.36*	1.60*	1.59*	1.61*
Inner London	0.30	0.28	−0.11	−0.17
Outer London	−0.08	−0.17	−0.30	−0.37
Metropolitan County	0.59	0.77	0.91	1.07
Nonmetropolitan Borough	0.15	0.19	0.16	0.08
Conservative Incumbent	−0.93*	−1.11*	−1.09*	−1.04*
Labour Incumbent	2.37*	2.95*	3.51*	3.97
Liberal Incumbent	−4.10*	−5.07*	−6.57*	−8.22*
SDP Incumbent	−1.21*	−1.48*	−2.01*	−2.59*
Liberal Candidate	−0.09	−0.15	−0.26	−0.43
R^2 (adj.)	0.90	0.90	0.89	0.89

All 20 regression coefficients for the variables representing aspects of the constituency electorate are statistically significant, and all have the expected signs. Labour won fewer votes, in all social classes, the more agricultural the constituency and the greater the percentage of the working class living in owner-occupied households: it won more votes the greater the proportions employed in energy and manufacturing industries and the higher the unemployment level. For all five variables, the relative size of the four coefficients was as expected as well; in constituencies with large numbers of miners, for example, Labour gained more votes from among the working than the middle class, and it lost most votes from among the working class in constituencies with large percentages of working-class owner-occupiers.

Compared to the results for the Conservative Party, the coefficients for the regional variables indicate a clear macro-spatial cleavage in voting Labour. In every region, Labour won more votes among all social classes than it did in the Southwest. In five, the differences were statistically significant; Labour had substantially greater support in the three northern regions and in both Greater London and East Anglia. Set against this, the urban:rural cleavage was much weaker for Labour. None of the coefficients is very large, and only for one constituency type – the metropolitan borough constituencies – did Labour do significantly better.

As with the Conservative Party, the Labour Party won more votes in marginal than in safe seats from all social classes, but in this case especially the working class; working-class voters were more loyal in constituencies where there was a greater chance of Labour losing. It also fared better where it was fielding an incumbent, and did significantly less well when its opponents – and especially the Liberal arm of the Alliance – was fielding an incumbent. Labour voting was not significantly influenced by which party represented the Alliance, although all four coefficients have negative signs suggesting that the Liberal candidates did slightly better than those of the SDP.

With regard to the *Alliance*, it is the regional pattern of voting that stands out in the regression results (Table 6.4). There is very little evidence of a local contagion or sectional effect, contrary to the findings of the Conservative and Labour analyses. The greater the proportion of social class II residents in the constituency, the greater the percentage of voters in social classes IV and V who supported the Alliance candidate, however, suggesting that only among the semi-skilled and unskilled manual workers was there continuity in voting for the 'third party'.

With regard to constituency socio-economic characteristics, Alliance candidates did significantly better among the working class in the more agricultural constituencies, suggesting that it

Table 6.4 Regression Analyses: Percentage in each Social Class Voting Alliance

Social Class	I and II	IIIN	IIIM	IV and V
Constant	30.47	27.09	26.92	25.80
Social Class II	−0.07	−0.03	0.05	0.10*
Agriculture	0.10	0.12	0.18*	0.24*
Energy	−0.12*	−0.14*	−0.19*	−0.23*
Manufacturing	−0.02	−0.03	−0.02	−0.03
Unemployment	−0.45*	−0.43*	−0.49*	−0.51*
Working Class O-O	−0.07*	−0.05	−0.03	−0.02
Marginality	0.04*	0.04*	0.04*	0.06*
North	−1.77*	−2.16*	−3.29*	−4.16*
Yorks/Humberside	−2.35*	−2.74*	−3.96*	−4.78*
Northwest	−3.38*	−3.56*	−4.53*	−5.20*
West Midlands	−3.50*	−3.44*	−4.07*	−4.44*
East Midlands	−4.16*	−4.04*	−4.64*	−4.94*
East Anglia	−3.90*	−3.87*	−4.46*	4.85*
Greater London	−6.94*	−6.89*	−8.21*	−8.80*
Southeast	−3.74*	−3.65*	−4.19*	−4.43*
Metropolitan Borough	−0.41*	−0.48	−0.77	−0.87
Inner London	0.04	−0.25	−1.00	−1.33
Outer London	1.53	1.41	1.46	1.42
Metropolitan County	0.40	0.31	0.09	−0.01
Nonmetropolitan Borough	−0.17	−0.18	−0.34	−0.40
Conservative Incumbent	−0.95*	−0.80	−0.68	−0.61
Labour Incumbent	−1.86*	−1.81*	−2.06*	−2.19*
Liberal Incumbent	15.68*	14.77*	14.78*	14.74*
SDP Incumbent	4.69*	4.32*	4.40*	4.35*
Liberal Candidate	0.87*	0.84*	0.95*	1.03*
R^2 (adj.)	0.49	0.54	0.63	0.69

replaced Labour as the destination of non-deferential working-class votes in rural England. The Alliance performed less well in areas of high unemployment and large percentages of energy workers, however, suggesting that voters in the areas of severest economic distress did not turn to the Alliance for possible salvation (Tables 6.2 and 6.3 suggest that Labour remained popular in these areas) and that the new force in British politics was substantially rejected in the stereotype Labour areas, the mining constituencies.

Voting for the Alliance displayed a major regional dimension, but with no accompanying urban:rural cleavage. The two parties polled significantly less than they did in the Southwest in every region, with Greater London and the Midlands standing out as the areas of poorest Alliance vote-winning across all classes, as did the northern three regions in the working class. Compared to the findings above relating to unemployment and mining, it is clear that the Alliance's inroads were very much confined to the parts of the country which lack either the traditional industries or the worst of the current recession.

The nature of the local contest was also of major importance in influencing the pattern of voting. Not surprisingly, Alliance incumbents, and especially Liberal incumbents, achieved percentages of the vote very substantially above the general level, in all social classes. Liberal candidates also outpolled SDP candidates in general (i.e. when incumbency was controlled for), by more than 1 per cent of the votes (an average of some 125 votes per constituency) among social classes IV and V. The presence of a Conservative incumbent did not influence the Alliance vote, except in social classes I and II. The presence of a Labour incumbent did, however, suggesting that in general terms the Alliance parties were better able to win votes from the Conservative Party in its strongholds than they were in the Labour heartlands; Labour voters, it seems, were wooed in the safe seats, for the less marginal the constituency the greater the percentage of all voters, but especially working-class voters, who supported Alliance candidates.

Together the results of these three sets of regressions confirm most of the hypotheses. They suggest that the Alliance did not produce a major realignment in English voting behaviour. The two established large parties continued to poll well in their areas of traditional strength, emphasising the continuity in the pattern of voting that has been present since the First World War (Johnston, 1983a). The Alliance made its greatest relative inroads in the seats that it already held, in the safe constituencies where protest voting was more likely to occur, and in two or three southern regions.

General Patterns and Unaccounted-for Variation

Comparison of the columns of each of the three tables (6.2-6.4) of results suggests very similar patterns; wherever the Conservative Party got a high percentage of the vote from one class it tended to get a high percentage from all others, for example. (This is in line with the 'imposed' results of Crewe and Payne's 1976 analyses.) The validity of this tentative interpretation was tested using principal components analysis. Every analysis contained four variables relating to the percentage of the members of each of the four classes estimated as voting for the relevant party (e.g. the four dependent variable data sets analysed in Table 6.2). The first component accounted for 99, 98 and 96 per cent of the variation for Conservative, Labour and Alliance respectively.

Such findings indicate that the pattern of geographical variation in voting in England in 1983 was very clear, and that where parties did well they did so among all classes. This, of course, is in line with the general thesis of this book, although the extent of uniformity in the patterns was somewhat surprising.

Given the commonality, it is sensible to combine the variables to portray both the geography of class voting and the geography of the residuals from the regression models. This was done using the principal component scores as the dependent variables in three further regression analyses. The results are given in Table 6.5, and confirm those in Tables 6.2-6.4. (The coefficients are much smaller, of course, because the dependent variables were measured in standard scores.)

The maps of the component scores (Figures 6.5-6.7) portray the overall geography of each party's support. Each geography is as expected. Virtually all of the major negative scores (those one standard deviation or more below the mean) for the Conservative Party are in the north of England (Figure 6.5) – with the only exceptions in Greater London, Bristol, Thurrock and parts of the West Midlands – whereas for Labour (Figure 6.6) there is a very solid block of above-average support in the north countered by below-average in the south: Labour did not appear to benefit from the poor Conservative performance in Greater London, however. Finally, the geography of Alliance voting was much 'patchier' (Figure 6.7); in general, its best performances were in the south and the worst in the north, however.

Table 6.5 Regression Analyses: Voting by Social Class –
Principal Components

	Conservative	Labour	Alliance
Constant	−0.45	−0.90	1.58
Class	0.10*	0.01*	0.04*
Agriculture	0.06*	−0.02*	−0.01
Energy	−0.03*	0.03*	−0.06*
Manufacturing	0.01	0.01*	−0.02*
Unemployment	−0.06*	0.02*	−0.06*
Working Class O-O	0.00	−0.01*	0.01
Marginality	−0.01*	−0.01*	0.05*
North	−0.23*	0.22*	−0.40
Yorks/Humberside	−0.30*	0.16*	−0.80*
Northwest	−0.16*	0.25*	−0.73*
West Midlands	0.08	0.04	−0.25
East Midlands	0.05	0.03	−0.55*
East Anglia	−0.06	0.10*	−0.89*
Greater London	−0.43	0.27*	−1.33*
Southeast	−0.05	0.06	−0.77*
Metropolitan Borough	−0.33*	0.08*	−0.09
Inner London	−0.40	0.01	−0.21
Outer London	−0.18	−0.01	0.28
Metropolitan County	−0.23	0.04	0.24
Nonmetropolitan Borough	−0.15*	0.01	0.18
Conservative Incumbent	0.25*	−0.06*	0.03
Labour Incumbent	−0.22*	0.17*	−0.28*
Liberal Incumbent	−0.31*	−0.31*	5.27*
SDP Incumbent	−0.07	−0.10*	0.78*
Liberal Candidate	0.01	−0.01	0.02*
R^2 (adj.)	0.92	0.90	0.61

Of more interest than the voting geographies set out in Figures 6.5–6.7, which are little more than alternative portrayals of the general distribution of votes across the country, are the maps of the residuals from regression. These identify the constituencies where the score for the relevant party is either substantially over- or underpredicted by the regressions in Table 6.5: they

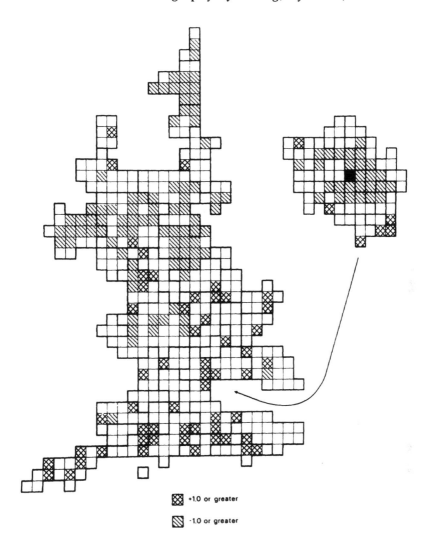

+1.0 or greater

-1.0 or greater

Figure 6.5 The Geography of Voting Conservative by Social Class: Scores on the Principal Component

show where the parties did better or worse than expected, given the characteristics and locations of the constituencies, and the nature of the local contest.

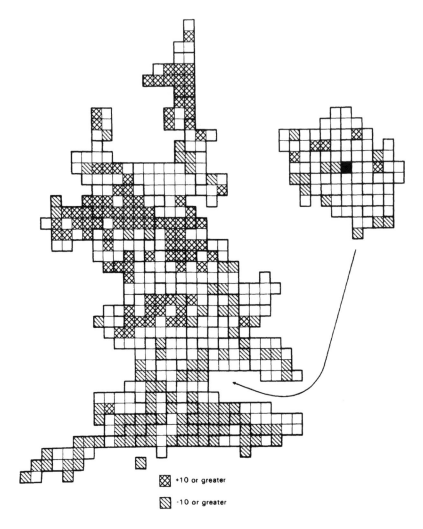

+10 or greater

-10 or greater

Figure 6.6 The Geography of Voting Labour by Social Class:
Scores on the Principal Component

The residuals from the regression for Conservative class voting
(Figure 6.8) identify several nodes of above-average support in
some urban constituencies (e.g. the Wirral, Bolton, Bury, north
Teesside, Walsall and Wolverhampton, parts of west and
southeast London, and much of Devon and coastal Dorset/
Hampshire). Countering these are pockets of less-than-expected
support, as in the Potteries, in Sheffield and neighbouring areas
of north Derbyshire in the Pennine valleys joining Lancashire to

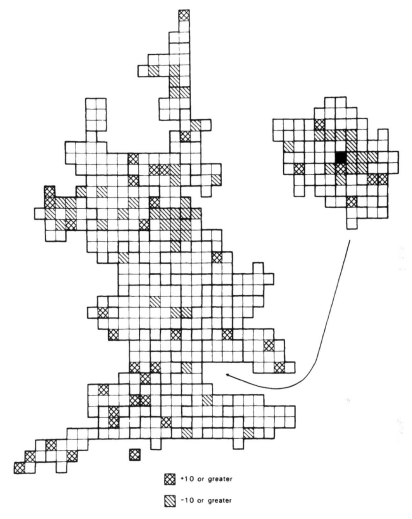

Figure 6.7 The Geography of Voting Alliance by Social Class: Scores on the Principal Component

Yorkshire, in Brent and Newham. Only the Sheffield and Potteries clusters suggest identifiable 'regions' where there is an apparent accentuation of anti-Conservative feeling, given the general characteristics of the constituencies; only the cluster in Wirral similarly identifies a 'region' of much higher-than-expected pro-Conservative attitudes. For the rest, local environmental circumstances appear to be the cause.

The residuals from the Labour regression (Figure 6.9) identify

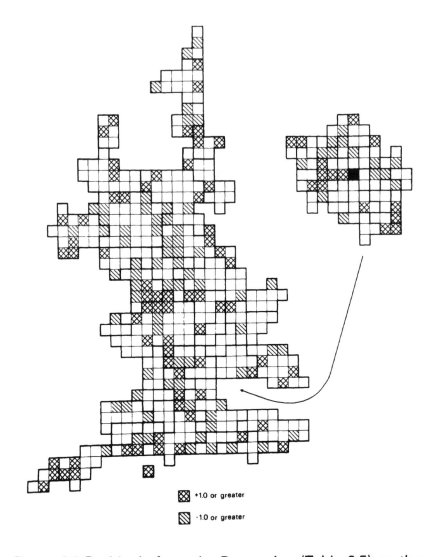

Figure 6.8 Residuals from the Regression (Table 6.5) on the Geography of Voting Conservative by Social Class

several major clusters of both above-average and below-average support for the party, given the overall pattern described by the regression model. Thus in Tyne and Wear, for example, Labour did better than expected in the 'inland' constituencies of Newcastle and Gateshead, but performed less well in the 'coastal' constituencies of Tynemouth, South Shields and Sunderland. Further south the Labour vote was substantially larger than that

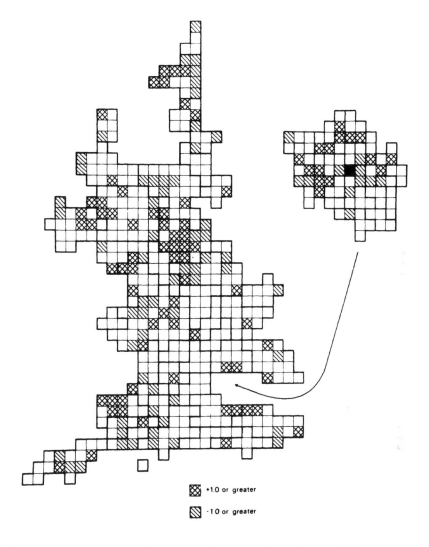

Figure 6.9 Residuals from the Regression (Table 6.5) on the Geography of Voting Labour by Social Class

estimated in Sheffield and adjacent constituencies, in the Potteries, and in west Lancashire, but substantially smaller in parts of Leeds and Bradford, and also in Nottingham and parts of the Black Country. In the south, Labour had three clusters of constituencies with better-than-expected support in north, east and west London, on the southern outskirts of Greater London (Dartford, East Surrey, Gravesham and Epsom) and in Bristol.

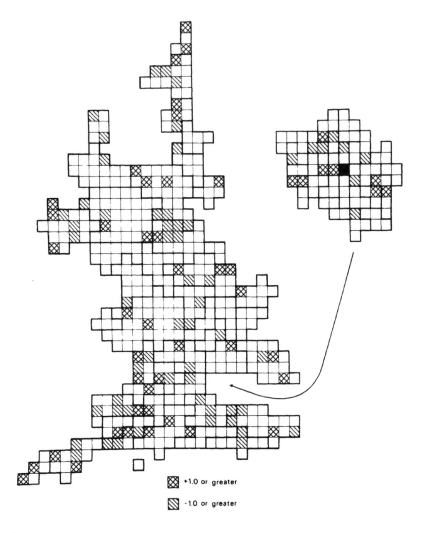

Figure 6.10 Residuals from the Regression (Table 6.5) on the Geography of Voting Alliance by Social Class

For the Alliance, the residuals also identify clusters of above- and below-average support (Figure 6.10). The former include parts of Durham, Crosby and Southport (the 'Shirley Williams effect'?), South Lincolnshire and four nodes within London. The latter include much of the Yorkshire coalfield, Brent in north London, several Surrey suburbs and the outskirts of Bristol.

Together, these three maps suggest that the regression model formulated and tested here did not exclude any major independent variables. The unaccounted-for variation can largely be attributed to local sub-regional nodes of support (and the lack of it), reflecting either particular local circumstances relevant to the 1983 election only or the continuation of long-established local patterns of voting which deviate somewhat from the regional trend.

And Those Who Did Not Vote?

So far, the geography of non-voting has been ignored in this analysis, largely because there has been much less study of the topic in general, on which valid hypotheses could be based. Complete absence of any analysis would leave an unfinished picture, however, so tentative explorations have been made.

The variable most likely to be associated with non-voting is marginality; the safer the seat, the greater the probability that people will be disinclined to vote. Associated with this, non-voting is likely to be greatest among those who are relatively isolated within their local environment (working-class Labour voters in middle-class areas, for example). For each of the four social classes, therefore, a class variable was introduced – middle-class owner-occupiers for classes I, II and IIIN; working class for IIIM, IV and V. It was anticipated that the percentage in a class not voting would be negatively associated with these variables, and positively with the marginality variable (which is the distance between the two leading parties in the estimated 1979 voting).

In addition to these two anticipated relationships, all of the variables used in the other models tested here were included to explore further the geography of non-voting. As with the other analyses, a general component was also derived from a principal components analysis in which the first component accounted for 91 per cent of the variation. Regressions for this component excluded the class variables.

The regression equations are given in Table 6.6. These show that the geography of non-voting, and especially that of non-voting by the non-manual classes, was very predictable. With regard to the expectations, the four coefficients for marginality are all positive and statistically significant: the safer the seat, the greater the probability that people did not vote. Regarding the influence of the class composition of the constituency, the larger the middle-class component of the electorate the smaller the percentages of middle-class voters who abstained; countering

Table 6.6 Regression Analyses: Non-voting by Social Class

Social Class	I & II	IIIN	IIIM	IV & V	Principal Component
Constant	21.44	242.24	19.80	25.22	−1.95
Class [+]	M.C.O-O −0.56*	−0.53*	W.C. 0.14*	0.08*	
Agriculture	−0.36*	−0.35*	−0.24*	−0.17*	−0.02
Energy	−0.03	−0.07	−0.15*	−0.23*	0.01
Manufacturing	−0.10*	−0.11*	−0.14*	−0.16*	−0.01
Unemployment	0.41*	0.38*	0.39*	0.31*	0.13*
Working Class O-O	0.04	0.05*	−0.02	0.03	−0.01
Marginality	0.14*	0.15*	0.15*	0.17*	0.03*
North	0.83	0.81	0.68	0.19	0.01
Yorks/Humberside	2.81*	2.87*	3.48*	3.01*	0.60*
Northwest	0.97	0.80	1.04	−0.13	0.10
West Midlands	1.25*	1.48*	2.46*	2.61*	0.26*
East Midlands	1.83*	2.03*	3.10*	3.22*	0.46*
East Anglia	1.80*	1.76*	2.11*	2.00*	0.41*
Greater London	4.69*	4.72*	6.51*	5.68*	1.29*
Southeast	1.90*	2.02*	2.95*	3.00*	0.53*

Metropolitan Borough	1.37*	1.36*	1.23*	0.95	0.27*
Inner London	4.75*	4.81*	5.26*	4.71*	1.12*
Outer London	0.67	0.64	-0.09	-0.13	-0.03
Metropolitan County	0.99*	1.01	0.78	0.50	0.10
Nonmetropolitan Borough	0.85*	0.96*	1.16*	1.17*	0.28*
Conservative Incumbent	-0.59	-0.50	-0.17	0.16	-0.11
Labour Incumbent	0.01	-0.18	-0.50	-0.89	0.05
Liberal Incumbent	-2.13*	-1.82*	-2.64*	-2.56*	-0.36
SDP Incumbent	-0.86	-0.20	-1.11	-1.01	-0.09
Liberal Candidate	-0.48*	-0.49*	-0.63*	-0.60*	-0.15*
R^2 (adj.)	0.82	0.78	0.66	0.59	-0.15

Note: †M.C. O-O – middle-class owner-occupier; W.C. – working class

this, the larger the working-class component, the larger the percentages of working-class voters who stayed at home. For the middle classes, it seems, local solidarity of views encourages voting, whereas for the working classes the opposite holds.

Turning to the variables relating to the constituency context, the higher the level of unemployment the greater the non-voting rate, presumably indicative of either alienation from the political process or of a 'plague on all of your houses' attitude. Non-voting was lower in the areas characterised by relatively high levels of both agricultural and manufacturing employment, however, and also in the mining areas, among the working class only (the last finding suggests that community solidarity in mining areas discouraged non-voting, in contrast to other working-class areas).

At a regional scale, the main findings relate to substantially higher than average levels of non-voting in Yorkshire and Humberside, in the East Midlands and in Greater London; perhaps even more noteworthy is that neither the North nor the Northwest differed significantly from the Southwest region. There was also a substantial urban:rural differential, with non-voting significantly larger in the urban constituencies, especially those in Greater London. (This may in part reflect population movements. The electoral roll used in June 1983 was compiled in October 1982. The inner cities could have lost population by migration in the interim period, and also above-average numbers by death.)

Apart from the finding regarding the marginality of seats, already noted, the significant relationships for the electoral context all refer to the Liberal Party; non-voting was significantly lower in areas with Liberal (rather than SDP) candidates, and especially in those with Liberal incumbents.

Although the principal components analysis showed that the four distributions were highly correlated, the regression with the component scores on the dependent variable has a relatively poor fit because in amalgamating the various classes a single independent variable representing class could not be included. Apart from this, the general pattern shown by the regression coefficients (the final column of Table 6.6) confirms the findings discussed above.

The map of the pattern of non-voting (Figure 6.11), using the scores on the principal component, shows considerable geographical clustering. High levels of non-voting were concentrated in Greater London, especially Inner London, and in a few urban

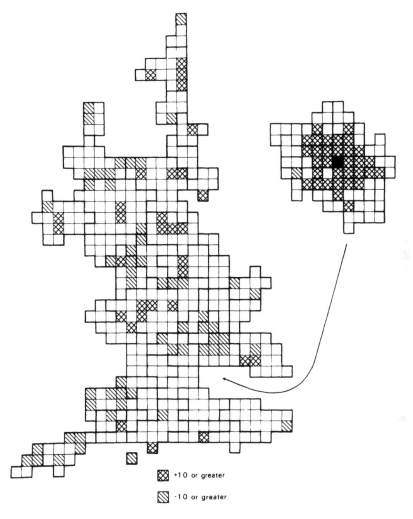

Figure 6.11 The Geography of Non-voting by Social Class:
Scores on the Principal Component

constituencies elsewhere (in parts of Birmingham, Sheffield,
Nottingham and Sunderland). Low levels of non-voting were
clustered in northeast Lancashire, the northern Home Counties,
Avon/Gloucestershire, Devon and Cornwall, and the Stafford-
shire/Derbyshire boundary area.

The regression reported in the final column of Table 6.6
accounts for two-thirds of this pattern. The residuals from this
(Figure 6.12) show where there are deviations from the general

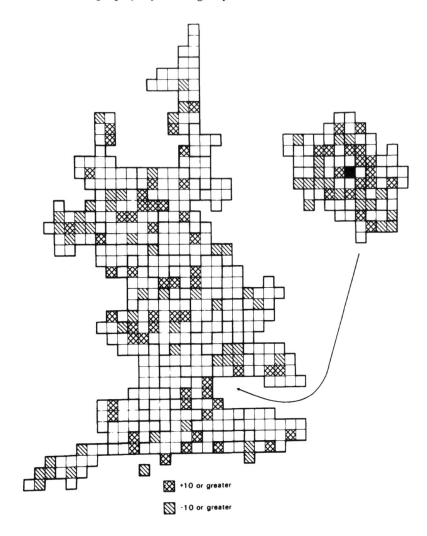

Figure 6.12 Residuals from the Regression (Table 6.6) on the Geography of Non-voting by Social Class

trend. Thus, for example, parts of Inner London had even larger non-voting proportions than estimated, whereas others (notably in the West and Southeast) had less. Within the Home Counties, a group of constituencies immediately to the west of London had substantially greater-than-expected non-voting, whereas in Hertfordshire there is a group where turnout was much greater than expected. Parts of Devon and Merseyside also had relatively high levels of turnout, suggestive of a high level of politicisation of the electorate. As with the other maps of residuals, therefore, Figure 6.12 picks out small areas of 'deviant' behaviour; further analyses are needed to indicate whether such behaviour has been consistent over a sequence of elections or whether it was particular to June 1983.

Social Class, Housing Tenure and the Geography of Voting

The analyses so far have investigated the 1983 geography of voting in England in the context of a single cleavage only, social class. Both theoretical and empirical work suggests the need to incorporate housing tenure as a further cleavage (see p. 15). Table 2.3 illustrates why. Using the traditional Alford (1963) index of class voting, reference to the top matrix there indicates a value of only 19.1, the difference between the percentage voting Labour of the lowest and that of the highest status group. If voter types are defined in terms of tenure as well as occupational class (the lower matrix in Table 2.3), the index increases to 24.2. The stereotypical non-Labour voter in 1983 was in a white-collar occupation and lived in an owner-occupier household; the stereotypical Labour voter was in a blue-collar occupation and lived in a rented dwelling.

To incorporate this extra cleavage, the matrix in the lower part of Table 2.3 was used in exactly the same way as already described, to produce estimates of the voting by members of the four groups shown there, in each of the 523 constituencies. The resulting frequency distributions are shown in Table 6.7 and in Figures 6.13-6.16. In general terms, what these show is very similar to that described for Table 6.1 and Figures 6.1-6.4: substantial inter-constituency variation in every one of the 16 cells of the matrix; and much greater variation in Labour voting than in either voting for the other two parties or non-voting.

To inquire into the possible determinants of these variations, the same regression models tested in the earlier part of this chapter were employed. In addition, principal components analyses were conducted, and accounted for 99, 99, 97 and 91 per cent respectively of the inter-constituency variation in Conservative, Labour and Alliance voting and non-voting. The scores on the principal components were also regressed against the same set of variables.

Table 6.7 Estimated Voting by Occupation and Housing Tenure: Summary Statistics

Type	1983 Vote	National Figure	Constituency Mean	Standard Deviation	Coefficient of Variation	Minimum	Maximum
White-collar Owner-occupier	Conservative	49.9	45.6	8.1	0.18	20	58
	Labour	10.7	13.5	7.1	0.53	1	34
	Alliance	20.7	19.5	4.9	0.25	7	44
	Did Not Vote	18.8	21.4	4.6	0.21	13	44
White-collar Renter	Conservative	29.5	31.3	7.6	0.24	11	44
	Labour	24.2	21.9	10.1	0.46	2	46
	Alliance	16.7	17.1	5.2	0.30	5	42
	Did Not Vote	29.6	29.7	5.0	0.17	20	51
Blue-collar Owner-occupier	Conservative	31.1	30.2	7.0	0.23	11	42
	Labour	19.1	18.6	8.9	0.48	2	40
	Alliance	20.8	20.1	5.7	0.28	6	47
	Did Not Vote	29.0	31.1	5.1	0.16	21	54
Blue-collar Renter	Conservative	14.4	16.5	5.0	0.30	5	36
	Labour	34.9	29.8	12.1	0.41	4	57
	Alliance	15.2	17.5	6.3	0.36	12	46
	Did Not Vote	34.5	36.2	5.7	0.10	22	62

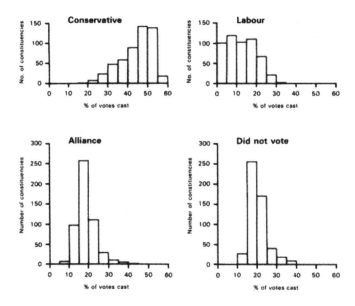

Figure 6.13 Frequency Distributions of Estimated Voting by Members of White-collar Owner-occupied Households

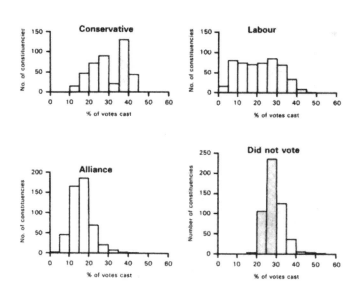

Figure 6.14 Frequency Distribution of Estimated Voting by Members of White-collar Tenant Households

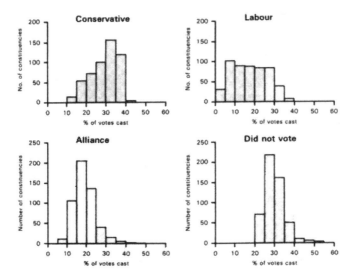

Figure 6.15 Frequency Distributions of Estimated Voting by Members of Blue-collar Owner-occupied Households

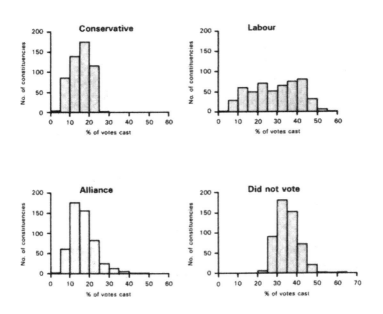

Figure 6.16 Frequency Distributions of Estimated Voting by Members of Blue-collar Tenant Households

The Results

The regressions for *Conservative* voting account for approximately five-sixths of the inter-constituency variation, with the significant regression coefficients emphasising the role of constituency charcteristics, the north:south and urban:rural cleavages and the impact of incumbency on the electoral outcome (Table 6.8).

Regarding constituency characteristics, most of the regression coefficients are as expected. Thus members of all groups were more likely to vote Conservative the larger the relative presence of middle-class owner-occupiers in the constituency, the greater the percentage of the workforce in agriculture, the smaller the percentage in energy industries, the lower the unemployment rate and the more marginal the seats (especially for owner-occupiers). More surprisingly, however, Conservative candidates picked up more votes where large percentages of the workforce are employed in manufacturing industries, and less in constituencies with relatively high percentages of working-class owner-occupiers. The former finding suggests, given that other variables such as percentage unemployed are held constant, that the Conservative Party was able to win above-average voter support in constituencies where jobs in manufacturing employment are still relatively plentiful (most of these are in the southern regions); the latter finding is less readily explained.

The findings relating to the other cleavages and to incumbency are very similar to those recorded in the earlier analysis of social class only (Table 6.2). Thus Conservative voters were significantly fewer in number in the three northern regions, but were well above-average in their presence among white-collar owner-occupiers in the West and East Midlands. Conservative voting was also smaller in proportion in all metropolitan and other urban constituencies, relative to the 'shire county' seats, although not significantly so in Greater London.

The residuals from the regression involving the scores – the Conservative principal component (the final column of Table 6.8) – are mapped in Figure 6.17. Many features of this map are common to that for the earlier analysis (Figure 6.8). There are several nodes of substantially larger-than-expected Conservative voting: on the Durham coast; in Blackpool and the Wirral; in parts of the north West Midlands; in southeast London; and around the Solent and Spithead. These are areas where members

Table 6.8 Regression Analyses: Percentage in Each Occupation/Tenure Group Voting Conservative

	White-collar Owner-occupier	White-collar Tenant	Blue-collar Owner-occupier	Blue-collar Tenant	Principal Component
Constant	43.79	30.93	29.49	16.35	-0.07
Middle Class O-O	0.63*	0.52*	0.49*	0.31*	0.07*
Agriculture	0.68*	0.67*	0.62*	0.47*	0.09*
Energy	0.68*	0.67*	0.62*	0.47*	0.09*
Manufacturing	0.06*	0.06*	0.07*	0.04*	0.01*
Unemployment	-0.33*	-0.38*	-0.33*	-0.25*	-0.05*
Working Class O-O	-0.11*	-0.11*	-0.10*	-0.07*	-0.01*
Marginality	-0.11*	-0.05*	-0.07*	0.00	-0.01*
North	-2.62*	-3.72*	-2.70*	-2.50*	-0.41*
Yorks/Humberside	-2.93*	-3.39*	-2.94*	-2.47*	-0.44*
Northwest	-1.39	-2.25*	-1.68*	-1.80*	-0.27*
West Midlands	1.83*	0.64	0.93	0.02	0.11
East Midlands	2.05*	0.83	1.12	0.18	0.14
East Anglia	0.58	-0.20	0.14	-0.38	-0.01
Greater London	-1.99	-3.07	-2.43	-2.19	-0.38
Southeast	0.90	0.16	0.35	-0.04	0.04

Metropolitan Borough	−2.53*	−2.35*	−2.23*	−1.45*	−0.31*
Inner London	−4.67	−4.00	−3.99	−2.51	−0.54
Outer London	−2.47	−2.06	−2.07	−1.35	−0.28
Metropolitan County	−2.03*	−1.90*	−1.81*	−1.21*	−0.25*
Nonmetropolitan Borough	−0.79	−0.89*	−0.85	−0.63*	−0.12
Conservative Incumbent	2.43*	2.02*	1.99*	1.22*	0.28*
Labour Incumbent	−0.67	−1.17*	−0.90*	−0.89*	−0.14*
Liberal Incumbent	−8.79*	−5.41*	−6.01*	−2.90*	−0.79*
SDP Incumbent	−2.16*	−1.12	−1.39	−0.50	−0.17
Liberal Candidate	−0.27	−0.04	−0.11	0.04	−0.01
R^2 (adj.)	0.81	0.85	0.83	0.86	0.84

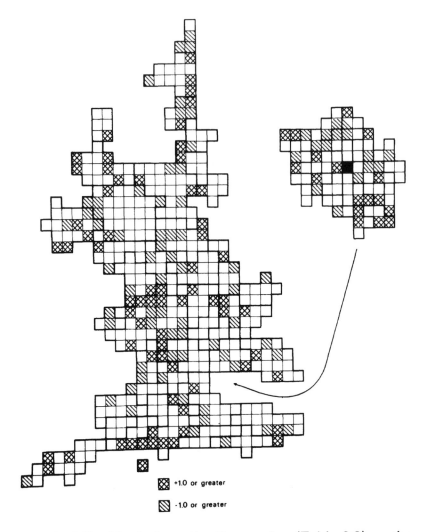

Figure 6.17 Residuals from the Regression (Table 6.8) on the Geography of Voting Conservative by Occupation and Housing Tenure

of all class-plus-tenure groups gave greater support to the Conservative Party than constituency characteristics and location suggest. Countering them are the nodes in west Lancashire, South Yorkshire and north Derbyshire, central Birmingham, northwest London and east London where the support for Conservative candidates was substantially less than in other comparable constituencies.

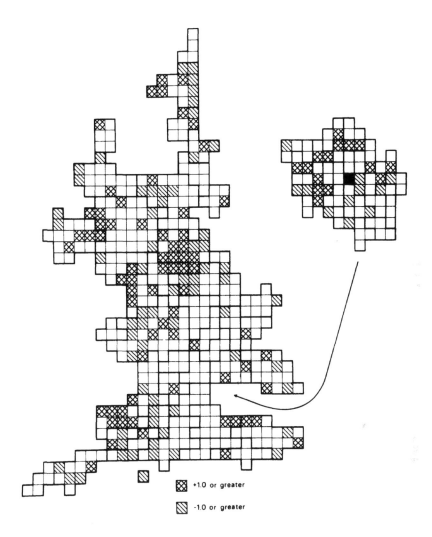

Figure 6.18 Residuals from the Regression (Table 6.9) on the Geography of Voting Labour by Occupation and Housing Tenure

Turning to the analyses of the *Labour* Party, the regressions also account for a little over five-sixths of the inter-constituency variation (Table 6.9). The pattern of coefficients is extremely similar to those (Table 6.3) for the analyses of voting by class alone, with one major exception; once tenure is included within the estimates, the negative relationship between voting Labour and the percentage of working-class households living in owner-

Table 6.9 Regression Analyses: Percentage in Each Occupation/Tenure Group Voting Labour

	White-collar Owner-occupier	White-collar Tenant	Blue-collar Owner-occupier	Blue-collar Tenant	Principal Component
Constant	0.24	5.87	4.27	13.43	-0.80
Working Class	0.10*	0.14*	0.12*	0.16*	0.01*
Agriculture	-0.27*	-0.43*	-0.35*	-0.53*	-0.02*
Energy	0.46*	0.63*	0.56*	0.72*	0.03*
Manufacturing	0.09*	0.14*	0.12*	0.18*	0.01*
Unemployment	0.39*	0.48*	0.43*	0.49*	0.02*
Working Class O-O	0.03	0.04	0.04	0.03	0.00
Marginality	-0.10*	-0.20*	-0.17*	-0.31*	-0.01*
North	3.31*	5.37*	4.56*	6.89*	0.26*
Yorks/Humberside	2.51*	3.79*	3.14*	4.67*	0.18
Northwest	3.83*	5.76*	5.04*	7.20*	0.28*
West Midlands	0.26	0.87	0.72	1.67	0.04
East Midlands	0.15	0.63	0.49	1.32	0.03
East Anglia	1.38	2.21*	1.87*	3.09*	0.11*
Greater London	4.51*	5.83*	5.13*	6.33	0.29*
Southeast	0.89	1.15	1.03	1.33	0.05

Metropolitan Borough	1.58*	1.76*	1.61*	1.47	0.09*
Inner London	0.48	0.31	0.20	-0.13	0.01
Outer London	0.03	-0.03	-0.03	-0.13	-0.01
Metropolitan County	0.87	1.06	0.94	0.98	0.05
Nonmetropolitan Borough	0.11	0.04	0.06	-0.17	0.01
Conservative Incumbent	-1.07*	-1.18*	-1.07*	-0.90	-0.06*
Labour Incumbent	2.67*	3.64*	3.34*	4.09*	0.18*
Liberal Incumbent	-4.58*	-6.73*	-6.23*	-8.92*	-0.34*
SDP Incumbent	-1.45*	-2.12*	-1.98*	-2.83*	-0.11*
Liberal Candidate	-0.20	-0.39	-0.30	-0.59	-0.01
R^2 (adj.)	0.86	0.87	0.85	0.87	0.86

occupied dwellings disappears. The Labour Party was strongest, among all groups of voters, in strongly working-class constituencies, in the northern regions and in Greater London, where it was fielding an incumbent, where unemployment was high, the seat was marginal and where agricultural workers were few but jobs in manufacturing and energy industries were relatively plentiful. And the residuals (Figure 6.18) highlight the areas where the party was unexpectedly relatively either very strong or very weak; the former include the Sheffield area, the west Lancashire coalfield, the Potteries, inner Birmingham, Bristol, parts of north and west London (again illustrating why so many of the regression coefficients for that area are statistically insignificant) and north Surrey; the latter include the Durham coast, Leeds-Bradford, Blackpool, Southport and Crosby, Nottingham, the northern Black Country and parts of the Dorset and Hampshire coasts.

For the *Alliance* (Table 6.10) it is once again the regional cleavage and the impact of incumbency that dominate the regression equations. The party did best among all types of voters where it was fielding a Liberal incumbent; SDP incumbents also did relatively well. Apart from incumbents, Liberal candidates tended to win about 1 per cent more of the votes than did their SDP counterparts. In most regions, the Alliance got 3-4 per cent less of the votes than it did in the Southwest, except in Greater London where its performance was twice as bad; it picked up more votes in the more agricultural and safer constituencies, but less in those with large percentages of miners and of unemployed. The residuals from the regression of the scores on the principal component (the final column of Table 6.10) identify several small nodes of relatively unexpected Alliance success – in parts of southwest (Richmond and Kingston), south-central (Bermondsey), southeast (Orpington) and inner-east (Bethnal Green, Bow areas where the Liberal Party retained relatively high levels of support throughout the 1920s) London; in much of west Devon and Cornwall; in Herefordshire; in Hertfordshire; and in south Lincolnshire (Figure 6.19) – plus others where its vote was substantially less than predicted – in much of Cumbria, for example; in several parts of north London; in the Croydon North West seat lost by the incumbent Liberal; in north Surrey; and in Leicester and Bristol.

Turning to the pattern of *non-voting* by occupation-plus-tenure groups, the regressions for individual groups each account for about two-thirds of the inter-constituency variation (Table 6.11); the statistical 'explanation' is much lower for the principal component because of the absence of a social class variable. (Nevertheless, the R^2 value of 0.44 for the general component is substantially lower than that of 0.67 reported for the analyses of non-voting by social class alone: Table 6.6.).

The regression coefficients suggest that a range of factors

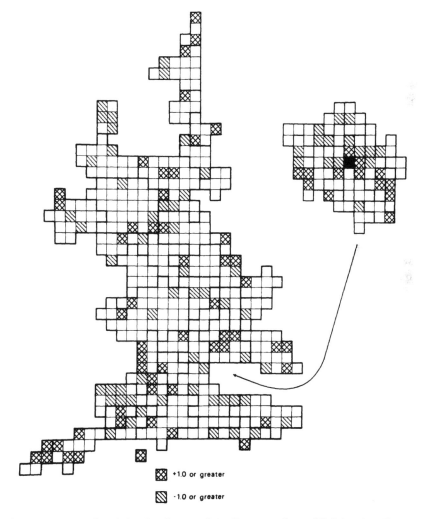

+1.0 or greater

-1.0 or greater

Figure 6.19 Residuals from the Regression (Table 6.10) on the Geography of Voting Alliance by Occupation and Housing Tenure

Table 6.10 Regression Analyses: Percentage in Each Occupation/Tenure Group Voting Alliance

	White-collar Owner-occupier	White-collar Tenant	Blue-collar Owner-occupier	Blue-collar Tenant	Principal Component
Constant	29.80	26.32	30.76	26.83	1.52
Social Class II	−0.05	−0.01	−0.01	0.04	0.01
Agriculture	0.07	0.17*	0.18*	0.30*	0.02*
Energy	−0.14*	−0.19*	−0.20*	−0.25*	−0.02*
Manufacturing	−0.02	−0.02	−0.02	−0.02	−0.01
Unemployment	−0.46*	−0.46*	−0.53*	−0.52*	−0.05*
Working Class O-O	−0.10*	−0.09*	−0.10*	−0.09*	−0.01*
Marginality	0.06*	0.06*	0.06*	0.07*	0.01*
North	−0.93*	−3.01*	−3.11*	−4.39*	−0.41*
Yorks/Humberside	−2.44*	−3.49*	−3.72*	−4.84*	−0.46*
Northwest	−3.40*	−4.10*	−4.40*	−5.23*	−0.51*
West Midlands	−3.49*	−3.64*	−4.02*	−4.30*	−0.42*
East Midlands	−4.11*	−4.16*	−4.63*	−4.79*	−0.48*
East Anglia	−3.84*	−4.04*	−4.43*	−4.76*	−0.47*
Greater London	−6.87*	−7.27*	−8.16*	−8.38*	−0.87*
Southeast	−3.71*	−3.75*	−4.18*	−4.22*	−0.44*

Metropolitan Borough	-0.52	-0.73	-0.84	-0.92	-0.10
Inner London	0.08	-0.67	-0.78	-1.33	-0.13
Outer London	1.43	1.17	1.34	1.06	0.12
Metropolitan County	0.23	0.06	0.07	-0.09	-0.01
Nonmetropolitan Borough	-0.22	-0.30	-0.37	-0.42	-0.03
Conservative Incumbent	-0.89	-0.68	-0.74	-0.59	-0.06
Labour Incumbent	-1.91*	-1.92*	-2.16*	-2.13*	-0.22*
Liberal Incumbent	15.58*	14.20*	15.43*	14.28*	1.58*
SDP Incumbent	4.58*	4.11*	4.60*	4.12*	0.45*
Liberal Candidate	0.91*	0.90*	0.99*	1.02*	0.10*
R^2 (adj.)	0.51	0.61	0.59	0.67	0.65

Table 6.11 Regression Analyses: Percentage in Each Occupation/Tenure Group Not Voting

	White-collar Owner-occupier	White-collar Tenant	Blue-collar Owner-occupier	Blue-collar Tenant	Principal Component
Constant	17.64	25.38	22.29	32.68	−0.20
Class†	M.C.O-O −0.24*	M.C.O-O −0.13	W.C. 0.02	W.C. −0.04	
Agriculture	−0.38*	−0.35*	−0.33*	−0.19*	−0.01*
Energy	−0.03	−0.15	−0.09	−0.27	−0.01
Manufacturing	−0.11*	−0.16*	−0.13*	−0.18*	−0.01*
Unemployment	0.39*	0.35*	0.49*	0.31*	0.01*
Working Class O-O	0.08*	0.07*	0.06	0.07*	0.01*
Marginality	0.14*	0.18*	0.17*	0.22*	0.01*
North	1.22*	0.91	1.07	−0.11	0.07*
Yorks/Humberside	3.18*	3.38*	3.78*	2.79*	0.12*
Northwest	1.33*	0.96	1.36*	0.03	0.06*
West Midlands	1.61*	2.30*	2.41*	2.62*	0.09*
East Midlands	2.10*	2.87*	3.13*	3.32*	0.10*
East Anglia	1.83*	2.02*	2.39*	2.02*	0.06*
Greater London	5.32*	5.55*	6.75*	5.14*	0.11*
Southeast	2.14*	2.66*	3.05*	3.08*	0.10*

Metropolitan Borough	1.32*	1.17*	1.34*	0.82	0.07*
Inner London	4.44*	4.67*	5.04*	4.24	-0.04*
Outer London	0.47	0.34	0.15	-0.03	0.07
Metropolitan County	0.83	0.63	0.61	0.15	0.02
Nonmetropolitan Borough	0.80*	1.06*	1.16*	1.22*	0.04*
Conservative Incumbent	-0.46	-0.15	-0.23	0.26	0.01
Labour Incumbent	-0.12	-0.59	-0.20	-1.01*	0.04*
Liberal Incumbent	-2.12*	-1.97*	-3.00*	-2.32	-0.08*
SDP Incumbent	-0.96	-0.84	-1.12	-0.72	-0.03
Liberal Candidate	-0.44*	-0.47	-0.62*	-0.48	-0.01
R² (Adj.)	0.71	0.64	0.64	0.66	0.44

Note: †M.C.O-O – middle-class owner-occupier; W.C. – working class

influenced the geography of non-voting. There were clear regional and urban:rural variations, for example, with turnout very low in Greater London (especially Inner London) and in the metropolitan counties. Almost every region had a significantly larger non-voting average than the Southwest, with the North and Northwestern regions somewhat surprisingly the main exceptions. As hypothesised earlier, non-voting increased the safer the seat; for every five percentage points increase in the gap between the leading two parties in 1979, non-voting among working-class tenants increased by 1.1 percentage points. In three of the groups, Liberal incumbents stimulated below-average levels of non-voting (the exception refers to working-class tenants), and among white-collar voters whatever their tenure, though especially among tenants, Liberal candidates for the Alliance were associated with significantly lower levels of non-voting than were SDP candidates.

All of the variables reflecting constituency characteristics were related to non-voting levels in at least some of the groups. In three cases, there were significant relationships in all four: non-voting tended to decrease as the percentage employed in agriculture increased, as the percentage employed in manufacturing increased, and as the percentage unemployed decreased. Among the working class, non-voting declined as the percentage employed in energy industries increased, whereas among owner-occupiers and working-class tenants, non-voting increased as the percentage of working-class owner-occupiers increased. Only the last of these is difficult to account for; for working-class owner-occupiers themselves, it implies that the conflict between class and tenure status led to abstention, but it is not clear why middle-class owner-occupiers were more likely to abstain in the presence of relative large numbers of working-class owner-occupiers.

Apart from this anomaly, the results of the regressions indicate that certain communities, mainly the rural, the mining and the still-prosperous manufacturing, have low levels of non-voting, perhaps indicative of a combination of community solidarity, political consciousness and a relative lack of alienation from the political system. Local economic distress appears to have stimulated such alienation, however, with non-voting greatest in the major metropolitan centres, in some (though not all) of the northern industrial regions, and in the constituencies with the highest levels of unemployment.

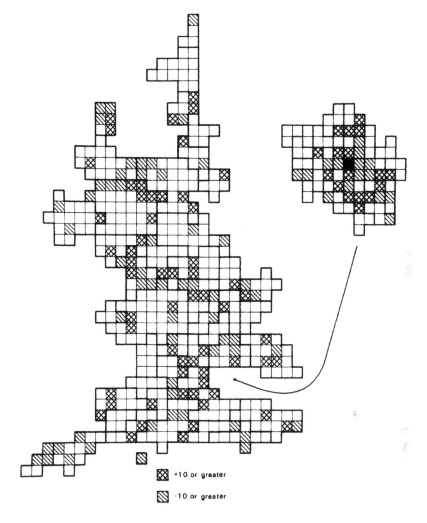

Figure 6.20 Residuals from the Regression (Table 6.11) on the Geography of Non-voting by Occupation and Housing Tenure

The residuals from the regression with the general component as the dependent variable suggest other, localised areas where non-voting was above or below expectations (Figure 6.20). Turnout was higher than expected in north Cumbria and parts of north Lancashire, for example, and also in parts of Staffordshire; it was substantially lower than expected (i.e. there were positive residuals, indicating high levels of non-voting) in the Potteries, in

Leicester and in Nottingham. Further south, non-voting was high immediately to the west of London, but low in parts of Devon and Cornwall. Within London itself, there was a complex pattern; non-voting was less than expected in the traditional East End constituencies, for example. (Non-voting was above-expected in Croydon North West, a seat lost by the Liberals and where incumbency had little impact: Figure 6.19.)

Where is Geographical Peculiarity Strongest?

In terms of goodness-of-fit, the various regression models tested in this chapter have been remarkably successful, suggesting that geographical variations in voting by socio-economic groups are very predictable. There are places where the predictions are less accurate, however; many of these have been attributed to local features.

Where, in more detail, are the major deviations from the general trends? Figure 6.21 summaries the previous four maps (Figures 6.17-6.20) by presenting a count of the number of times each constituency falls outside the one standard error band of the regression equations. In this, a value of 0 indicates a constituency for which all four predictions (percentage Conservative, Labour, Alliance, non-voting) were relatively accurate, whereas a value of 4 indicates a constituency where all the predictions were way out of line. Needless to say there are many more (177) in the former than in the latter category. In only eight constituencies were all four predictions very wrong, indicating a very unexpected distribution of the votes: in Bethnal Green and in Chelmsford the Alliance vote was much larger than anticipated, whereas all of the other three were much smaller; in Normanton and in Bristol West it was the Labour and non-voting percentages that were well-above expected; in Dewsbury, Romsey and the Isle of Wight both the Conservative and the Alliance votes were relatively large while Labour and non-voting were small; and in Barking only the Labour vote was substantially overpredicted, whereas all the other three were underpredicted.

The general impression of the map is that there were several blocks of constituencies which conformed very much to the general trends. Much of Norfolk and Suffolk fall in this category, with only Ipswich (a rather surprising Labour retention), and Northwest Norfolk (contested by the Conservative defector to

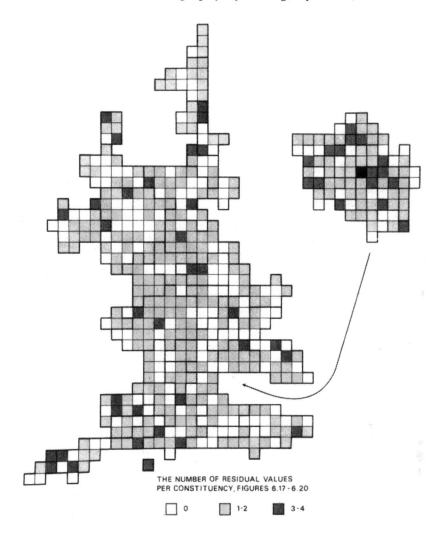

THE NUMBER OF RESIDUAL VALUES
PER CONSTITUENCY, FIGURES 6.17 - 6.20

☐ 0   ▨ 1-2   ■ 3-4

Figure 6.21 The Geography of 'Deviant' Voting: the Number of Residuals per Constituency Reported in the Maps of Figures 6.17-6.20

the SDP, Christopher Brocklebank-Fowler) recording two substantial residuals. The nearby area comprising the semi-rural East Midland constituencies of Gedling, Rushcliffe, Loughborough, Stamford, Rutland, Lincoln, Newark and Sherwood stand out as non-deviants, as does the block further south comprising Blaby, South Leicester, Harborough and the two Northampton seats. Similarly the two Dudley seats and Halesowen and the

Worcestershire seats are 'conformist areas', along with much of west Kent, the western part of Greater Manchester and north Humberside. Countering this there are areas containing many deviations. Outstanding among these are parts of Greater London, indicating the variability of response to the election there. No blocks of contiguous 'conforming' constituencies stand out in London, but there are several where the regressions were clearly relatively unsuccessful: in Brent, Barnet, Ealing, Richmond, Sutton, Bow, Newham, Leyton, Lewisham and Orpington. Elsewhere, in parts of the West Country (East and North Cornwall, Teignbridge and Plymouth Devonport – all constituencies with greater than expected Alliance success), in Bournemouth, Poole, Gosport and Havant (in all of which the Labour vote collapsed), over much of the Midlands, in the Sheffield, Potteries (both good for Labour), Nottingham (bad for Labour) and Leicester (bad for the Alliance) areas, and in much of coastal Northumberland and Durham, plus Cumbria, for example, are blocks of constituencies where the predictions were less successful. These are the parts of England where either well-established party loyalties or particular aspects of the local campaign in 1983 produced results that were somewhat out of line with those for similar constituencies, similarly located.

In Summary

The analyses reported here have provided a broad measure of support for the various hypotheses outlined in Chapter 4 and at the start of this chapter. There were substantial inter-constituency variations in the support given to the parties (especially Labour) by both social class and occupation/tenure groups, and most of these variations could be successfully accounted for in the regression analyses. 'Who voted what, where?' in 1983 was a function not only of their socio-economic characteristics but also the characteristics of their neighbours, the local electoral context and the location of their constituency.

Fuller consideration of these results is presented in Chapter 10. At this stage, the simple conclusion is drawn that any understanding of the 1983 election in England must include a substantial geographical component. What has yet to be established is whether this represents major geographical changes since the previous election.

7 THE GEOGRAPHY OF CHANGING VOTER ALLEGIANCE

The previous chapter has focused on the geography of voting in 1983, showing substantial variations in the support for each party from various class, occupation and tenure groups. The present chapter complements those findings with analyses of the shifting pattern of votes, focusing on spatial variations in the estimated values of the constituency flow-of-the-vote matrices. It seeks to establish whether changes in the pattern of voting between 1979 and 1983 were the same in all parts of England or whether different places recorded different shifts.

The Pattern of Shifts

The national flow-of-the-vote matrix, after smoothing, is shown in Table 2.4B. The percentages shown there were applied to the 1979 distribution of votes, to produce the gross flow matrix shown in Table 7.1. This provided one of the sets of constraints (the matrix V) for the entropy-maximising procedure; the other sets of constraints were the matrices showing the estimated results in the 523 constituencies in 1979 (BBC/ITN, 1983) and the 1983 results.

Table 7.1 The Gross Flow of Votes, 1979-83

1983 Vote 1979 Vote	Conservative	Labour	Alliance	Did Not Vote
Conservative	8,955,448	421,752	1,362,943	1,380,753
Labour	653,074	5,328,510	1,934,327	1,292,202
Liberal	542,116	315,941	2,513,233	383,089
Did Not Vote	1,542,017	773,416	913,631	6,781,415

Only six iterations were needed to give a set of 523 constituency flow-of-the-vote matrices whose sums were within an average of 5 per cent of the relevant constraint values. The selected set had an average deviation of 4.0 per cent from the 1979 election results, an average deviation of only 0.08 per cent from the 1983 results, and fitted the national flow-of-the-vote matrix exactly.

For eight of the constituencies (Birmingham, Sparkbrook; Dudley, West; Knowsley, North; Lancashire, West; Salford, East; Walsall, South; Warley, East; West Bromwich, West) the absence of Liberal candidates in 1979 precluded any estimate of the Liberal vote there. Thus for these eight, the values in the third row of the matrix were all zero. As a consequence, the values in the other three rows are all substantially different (when expressed as percentages of the row totals) from those in the other 515 constituencies. These eight are excluded from all of the analyses in this chapter.

To identify the amount of inter-constituency variation, the cell values in each matrix were expressed as percentages of the relevant row totals. The summary statistics for each of the 16 cells, along with the national values, are given in Table 7.2. Frequency distributions are given in Figures 7.1-7.5.

The distributions of estimated party loyalty by constituency (Figure 7.1) show a clear difference betwen that for Labour and those for the other parties and for non-voting. Variability around the Conservative mean of 72.6 per cent was small, and almost two-thirds of the constituencies fell into the modal category. The same is true for Non-voting. For the Alliance there was slightly more variability (the coefficient of variation — SD/Mean — is 0.11), but almost all of the constituencies fell within a 30 percentage point range. With the Labour Party, however, spatial variability in party loyalty was very much greater (a coefficient of variation of 0.24). The distribution was also negatively skewed, indicating a substantial number of constituencies (about 20 per cent of the total) in which the party retained less than 40 per cent of its 1979 supporters. Each of the other distributions has a slight negative skewness, but none as pronounced as that for Labour; the latter's failure to hold voter loyalty in such a large percentage of the constituencies is the most striking element of the distributions.

As with the pattern oi party loyalty it is the distributions of flows involving the Labour Party which in general show the greatest variability. This comes through especially in the

Table 7.2 The Constituency Flow-of-the-Vote Matrices: Summary Statistics

1979	1983	National	Mean	SD	CV	Minimum	Maximum
Conservative	Conservative	73.9	72.9	5.8	0.08	37	84
Labour	Labour	57.9	53.4	12.7	0.13	20	77
Liberal	Alliance	66.7	65.0	6.9	0.11	32	84
Did Not Vote	Did Not Vote	67.5	67.5	4.9	0.07	48	84
Labour	Conservative	7.1	8.1	2.9	0.36	2	16
Liberal	Conservative	14.4	14.2	2.7	0.19	3	22
Did Not Vote	Conservative	15.5	15.7	3.9	0.25	5	31
Conservative	Labour	3.5	4.0	2.1	0.53	1	11
Liberal	Labour	8.4	10.0	4.9	0.49	2	27
Did Not Vote	Labour	7.8	7.7	3.0	0.39	2	16
Conservative	Alliance	11.2	11.3	2.7	0.24	4	37
Labour	Alliance	21.0	23.6	8.4	0.36	5	49
Did Not Vote	Alliance	9.2	9.2	2.6	0.28	2	21
Conservative	Did Not Vote	11.4	12.1	3.5	0.29	5	28
Labour	Did Not Vote	14.0	14.9	3.4	0.23	6	27
Liberal	Did Not Vote	10.2	10.8	2.9	0.27	4	24

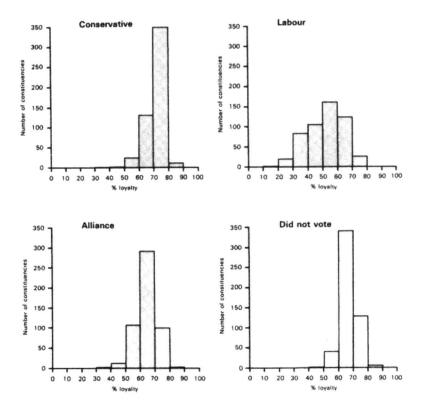

Figure 7.1 Frequency Distributions of Percentages Loyal to Each Party, by Constituency, 1979-83

movement from voting Labour in 1979 to Alliance in 1983, in which the distribution is very platykurtic (Figure 7.4) compared to almost all of the others. In all cases, however, the geographical variation was quite substantial: no coefficient of variation is less than 0.08, and many are larger than 0.20. There was, then, substantial spatial variability in the flow-of-the-vote between 1979 and 1983.

What of the geography of these variations? Maps of those constituencies more than one standard deviation from the mean indicate very clear spatial patterns, in almost every case. Figure 7.6, for example, maps the geography of loyalty to the Conservative Party. Very few constituencies in the north of England had more than 78.3 per cent loyalty (Barrow; Westmorland; Boothferry; Ribble Valley; Glanford and Scunthorpe;

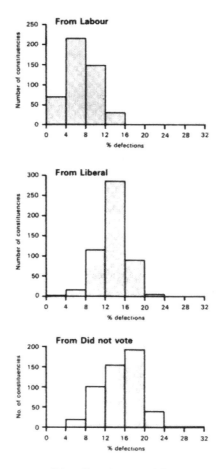

Figure 7.2 Frequency Distributions of Percentage Defections to Conservative, by Constituency, 1979-83

and West Derbyshire): much of the Northern region, of South Yorkshire, of Greater Manchester and of Merseyside, on the other hand, had loyalty rates of less than 66.8 per cent. Further south, the picture was reversed. Apart from 22 constituencies in London, there were only four (the Ladywood and Small Heath constituencies in Birmingham; Coventry, North West; and Plymouth, Devonport) which recorded substantially below-average levels of Conservative loyalty. Very high levels of loyalty to that party were also relatively rare in the southern regions (indicative of the skewed distribution: Figure 7.1) and there was none in Greater London.

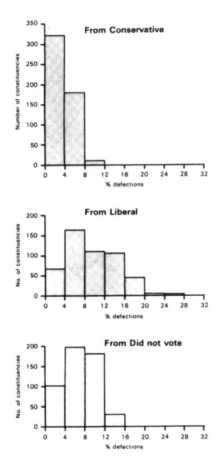

Figure 7.3 Frequency Distributions of Percentage Defections to Labour, by Constituency, 1979-83

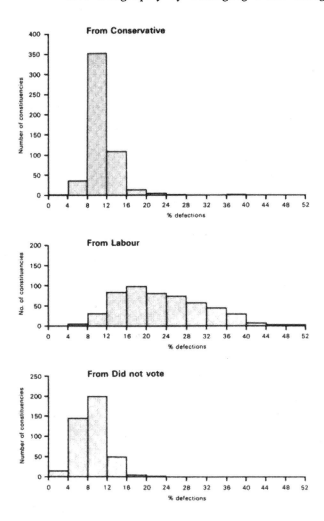

Figure 7.4 Frequency Distributions of Percentage Defections to Alliance, by Constituency, 1979-83

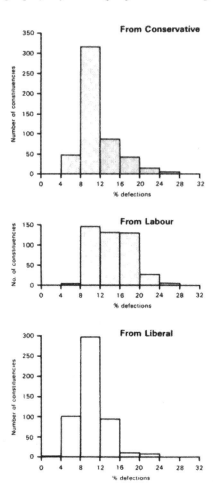

Figure 7.5 Frequency Distributions of Percentage Defections to Non-voting, by Constituency, 1979-83

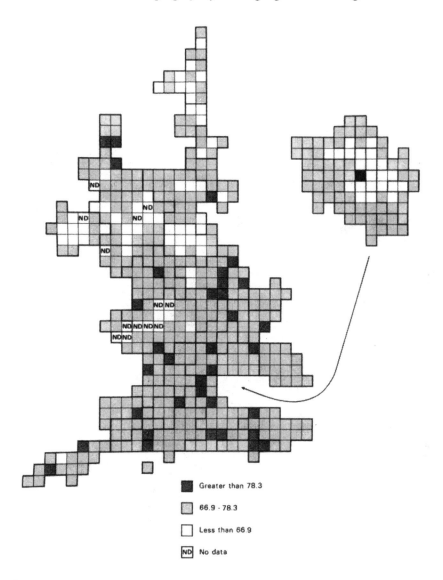

Figure 7.6 Loyalty to the Conservative Party, 1979-83

On the map of Labour loyalty (Figure 7.7), the pattern is almost the reverse of that for the Conservative Party. In the north, only a small number of either rural or high-status suburban constituencies recorded Labour loyalty rates of less than 40.8 per cent; they included Harrogate, Richmond and Ryedale in Yorkshire, Fylde in Lancashire and Crosby and

Figure 7.7 Loyalty to the Labour Party, 1979-83

Southport in Merseyside. Much of the north, and also of the West Midlands and Inner London, contained constituencies with more than 66.1 per cent loyalty among Labour's 1979 voters. The southern Home Counties and nearly all south-coast constituencies, on the other hand, form a solid block of substantial defections from the Labour ranks; only in Ipswich, Oxford East

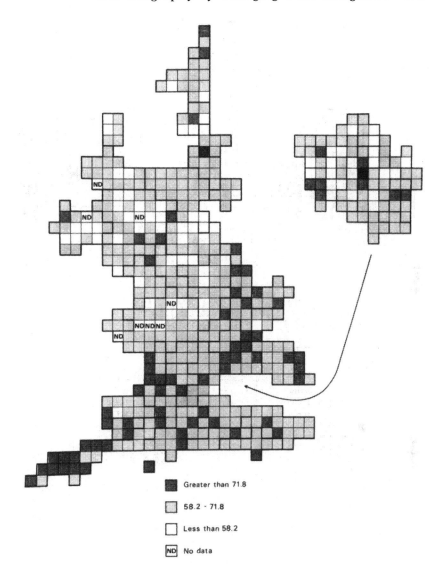

Figure 7.8 Loyalty to the Alliance Parties, 1979-83

and Slough was the estimated loyalty above two-thirds of the 1979 Labour votes.

Figures 7.6 and 7.7 indicate the well-known north:south split in England, with the Conservative heartland in the south (excluding parts of Inner London) and the Labour-dominated constituencies in the north (except some of the more affluent

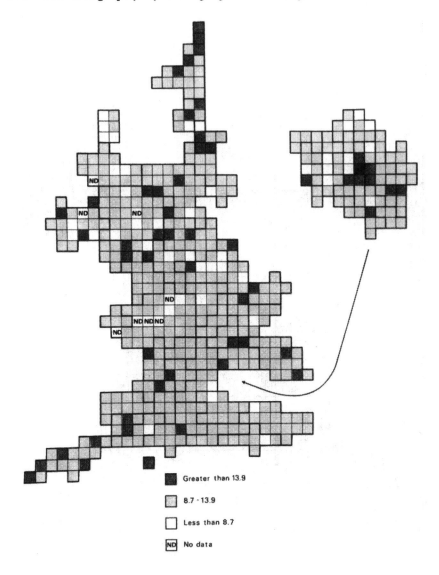

Greater than 13.9

8.7 - 13.9

Less than 8.7

No data

Figure 7.9 Flows from Conservative to Alliance, 1979-83

suburban and rural areas). The Alliance map (Figure 7.8) reproduces much of this. There are few constituencies in the north (Berwick; Blyth Valley; Durham; Stockton South; Colne Valley; Congleton; Hazel Grove; Cheadle; and Crosby) where the Alliance held on to more than 71.8 per cent of the 1979 Liberal vote (most of these are constituencies where the Liberal

Greater than 31.9

15.3 - 31.9

Less than 15.3

ND No data

Figure 7.10 Flows from Labour to Alliance, 1979-83

Party traditionally performs well. Stockton South was won by the SDP). Further south, on the other hand, only in Ipswich, Slough and Bristol South (plus 21 constituencies in Greater London) did the Alliance candidates fail to retain at least 58.2 per cent of the 1979 Liberal support; in Devon and Cornwall it held more than 71.9 per cent in almost every constituency.

Figure 7.11 Flows from Non-voting to Alliance, 1979-83

Maps of all 13 other cells of the flow-of-the-vote matrix are not given here, but Figures 7.9-7.11 show the geography of the flow to the Alliance. That for the Conservative-Alliance flows (Figure 7.9) suggests that the new party grouping was most succesful in attracting Conservative support (13.9 per cent or more) in the Liberal heartland of the Southwest, and also in Inner London and north of England seats where the Conservative party is generally weak. It was best able to attract Labour votes (31.9 per cent or more) in the South (Figure 7.10), performing above-average in only a few small pockets in the North (the Crosby, Stockton and Cheadle-Hazel Grove effects). Similarly, the Alliance was most successful in winning over 1979 non-voters in southern constituencies, excluding most of those in Inner London (Figure 7.11).

Regression Analyses

For these analyses, the 16 cells of the flow-of-the-vote matrix form the dependent variables. The independent variables, and the hypotheses to which they relate, are as follows:

Conservative 1979; Labour 1979; Liberal 1979.

It is hypothesised that the larger the party's percentage of the vote in a constituency in 1979, the greater its ability to retain the loyalty of its supporters then and also to win over those who either voted for other parties or who did not vote in 1979.

Regional Variations

In general, the Labour Party is expected to retain party support and to win converts in the northern regions, whereas the Conservative Party and the Alliance are expected to do better in the south. (Full descriptions of the regions are in Chapter 6 and Appendix 2.)

Type of Constituency

A sixfold classification is used, as before:

Metropolitan Borough   Metropolitan County
Inner London   Nonmetropolitan Borough
Outer London   Nonmetropolitan County

This reflects the local government division of the country — into the metropolitan and nonmetropolitan counties, plus Greater London; a division of London into inner (the ILEA boroughs) and outer; and the Boundary Commission's division of constituencies into borough (i.e. entirely urban) and county. In total, this classification provides a rough urban:rural and inner city:suburban division. Labour is expected to do better in the more urban classes.

Incumbency

Incumbency is expected to bring advantage to a candidate, so where an incumbent was standing (see p. 118) the relevant party should both retain above-average loyalty and win more converts from other parties. Within the Alliance, Liberal candidates might perform better than SDP candidates.

Marginality

Voters are expected to be less likely to switch between parties in the more marginal constituencies. Marginality was measured as the difference in percentage points between the leading two parties in the 1979 estimated results: thus there should be a negative relationship between this measure and the degree of intra-party loyalty.

Embourgeoisement and Depression

During the period of the 1979-83 Conservative government, unemployment increased substantially and increased numbers of working-class householders were able to become owner-occupiers. High levels of unemployment should be associated with shifts away from the Conservative Party; high percentages of working-class owner-occupiers should be associated with above-average loyalty and shifts to that party. Data on changes between 1979 and 1983, at the constituency level, are not available. The 1981 values (percentage unemployed; percentage working-class residents in owner-occupied homes) are used, on the assumption that the relative positions of the constituencies on these did not vary between 1979 and 1983.

In all, 22 independent variables are included in the regression equations. To counter the effects of such a large number of such variables, the goodness-of-fit measure, R^2, has been adjusted to allow for the decline in the degrees of freedom (Nie *et al.*, 1975; Draper and Smith 1966).

The Results

The results of the twelve regressions for flows terminating in one of the three parties (i.e. the cells of the first three columns of Table 7.1) are given in Tables 7.3-7.5. Overall, they indicate a very high level of predictability. The average adjusted R^2 value of 0.65 suggests that two-thirds of the pattern of flows can be accounted for. Perhaps not surprisingly, the lowest correlations refer to voters for the Alliance in 1983 (average R^2, 0.47); countering this, the pattern of shifts to Labour was the most predictable (average R^2, 0.81).

Looking first at the four regressions relating to the *Conservative* vote in 1983 (Table 7.3), the general impression is of continuity: the Conservative Party did well in 1983 in the same places as 1979. There are, however, some major differences between the four equations.

For the pattern of Conservative loyalty (column 1), the major influence was clearly Conservative strength in 1979 (exactly half of the variation in the dependent variable is accounted for by that one variable alone). Apart from this, Conservative loyalty was greatest in the nonmetropolitan, county constituencies — suggesting a clear urban:rural split in the party's support: further, it was substantially eroded in constituencies fought by Alliance incumbents; was greater in the more marginal constituencies; and fell by one percentage point with every four percentage points increase in the level of unemployment. Only in the Northwest was there a significantly below-average level of loyalty at the regional scale.

Regarding the flow of votes from Labour to Conservative, apart from the general correlation with Conservative strength in 1979 the major finding is the strong inter-regional as well as urban:rural differentiation in the size of the flow. The Conservative Party was apparently better able to win voters from Labour in the south of the country than in the north, with the Northwest region having the smallest percentages apparently willing to transfer their allegiance; it also won substantially fewer in the rural than the urban constituencies, but more in the suburban than in the inner city seats. In addition, Labour incumbents lost less votes to the Conservative candidates than did non-incumbents, as did all Labour candidates in marginal seats (the flow was significantly greater in the safer constituencies) and those in constituencies where unemployment was high.

Very similar patterns are revealed by the regression of the former non-voters who shifted to Conservative. These, too, were

Table 7.3 Regression Analyses: Flow-of-the-Vote to Conservative

1979 Vote 1983 Vote	Conservative Conservative	Labour Conservative	Liberal Conservative	Did Not Vote Conservative
Constant	64.46	4.83	9.36	17.19
Conservative 1979	0.25*	0.10*	0.12*	0.10*
North	-0.54	-1.30*	0.50	-1.49*
Yorks/Humberside	-0.69	-0.98*	0.70	-1.47*
Northwest	-1.47*	-1.52*	0.49	-1.70*
West Midlands	-0.11	-0.90*	0.66	-0.42
East Midlands	-0.17	-0.93*	1.29*	-1.21*
East Anglia	0.09	-0.31	0.78	-0.30
Greater London	-1.41	-0.67	1.26	-2.41
Southeast	-0.61	-0.50*	0.48	-1.06*
Metropolitan Borough	-3.34*	-1.48*	-1.64*	-2.89*
Inner London	-6.86*	-2.49*	-2.63*	-4.60*
Outer London	-2.38	-1.37	-1.68	-2.50
Metropolitan County	-1.71*	-1.27*	-0.88	-2.14*
Nonmetropolitan Borough	-1.60*	-0.93*	-0.72*	-1.76*

Conservative Incumbent	0.11	0.10	0.19	0.02
Labour Incumbent	0.38	−0.39*	0.68*	−0.43
Liberal Incumbent	−5.21*	−0.12	−3.30*	−1.47*
SDP Incumbent	−3.14*	−0.19	−2.53*	−0.50
Liberal Candidate	−0.26	0.08	−0.22	0.08
Marginality	−0.09*	0.04*	−0.03*	−0.05*
Unemployment	−0.27*	−0.14*	−0.04	−0.28*
Working Class O-O	0.05	0.01	0.02	0.01
R² (adj.)	0.71	0.81	0.49	0.71

concentrated in southern, nonmetropolitan, non-urban constituencies, where unemployment was low, and the Conservative Party already strong. However, if the seat was marginal, the flow was significantly above-average.

The flow of 1979 Liberal voters to the Conservative Party in 1983 was much less predictable. It was greatest in those constituencies where the recipient party was strong in 1979, and was even greater in the East Midlands, but much reduced in the borough constituencies and in Inner London. The presence of an Alliance incumbent candidate substantially reduced the flow.

Overall, then, the Conservative Party showed continuity of support. It was especially strong in rural England, and in the more prosperous constituencies with low levels of unemployment. In winning support from Labour, and also from non-voters, it was least successful in the northern regions. Yet there were areas in the north where it performed relatively well. The map of the major residuals (1.0 SE or greater) for the regression on Conservative loyalty (Figure 7.12) shows a number of northern constituencies, including some in urban areas (such as Tyneside, Hull and Bolton), where loyalty to the party was underpredicted by at least three percentage points; countering this, its performance was substantially overpredicted in parts of South Yorkshire.

The regressions for the *Labour* Party (Table 7.4) illustrate its strength in its traditional heartlands and its appeal to the relatively underprivileged. Thus, in the regression for Labour loyalty, the first coefficient shows that every increase of the Labour vote in 1979 by two percentage points produced about one percentage point more loyal voters; a three-percentage-point increase in unemployment produced a similar result. There was also a major north:south split, with Labour loyalty on average eight percentage points higher in the Northwest than in the Southwest. Labour incumbents, too, retained above-average proportions of the 1979 vote; Liberal and, especially, the SDP incumbents (all but one of the latter were formerly Labour Party MPs) eroded the Labour vote, however. Loyalty was also much greater in the more marginal seats.

Flows to the Labour Party were very predictable, and also very closely related to where the party was strong in 1979. (Regressions on the first independent variable alone accounted for 75 per cent of the Conservative-Labour flow, 68 per cent of

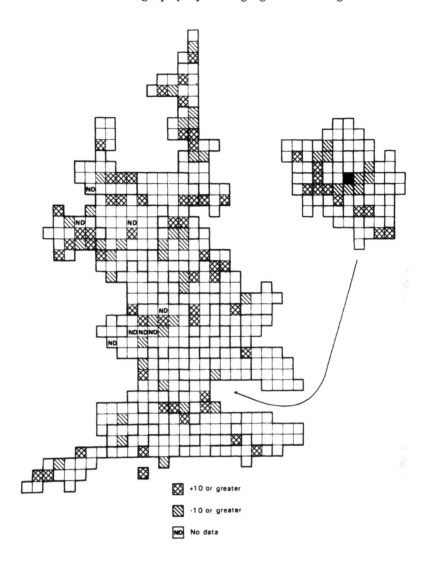

Figure 7.12 Residuals from the Regression on Conservative Party Loyalty

the Liberal-Labour flow and 65 per cent of the Non-voters-Labour flow.) The Northwest stands out as the region where Labour was most successful in winning converts.

Four other regression coefficients are particularly worthy of mention in Table 7.4. Two of the four relating to the Alliance candidates are significant. Former Conservative voters and non-

Table 7.4 Regression Analyses: Flow-of-the-Vote to Labour

1979 Vote 1983 Vote	Labour Labour	Conservative Labour	Liberal Labour	Did Not Vote Labour
Constant	33.89	−0.25	1.53	4.44
Labour 1979	0.46*	0.08*	0.15*	0.10*
North	5.74*	0.29	1.02	1.02*
Yorks/Humberside	4.89*	0.31	0.97	0.54
Northwest	8.01*	0.95*	2.75*	1.58*
West Midlands	5.20*	0.25	1.12*	0.95*
East Midlands	5.35*	0.30	1.72*	0.57*
East Anglia	1.81	−0.07	0.39	−0.04
Greater London	1.98	0.06	1.27	−0.61
Southeast	1.70	0.06	0.66	−0.04
Metropolitan Borough	2.19*	0.67*	1.11*	0.41
Inner London	4.51	1.04	1.30	0.37
Outer London	1.48	0.27	0.15	0.13
Metropolitan County	1.31	0.12	0.56	0.06
Nonmetropolitan Borough	0.44	0.03	−0.02	−0.04

Conservative Incumbent	0.62	-0.13	-0.26	-0.19
Labour Incumbent	2.42*	0.44*	1.72*	0.81*
Liberal Incumbent	-4.94*	0.17	-2.61*	0.13
SDP Incumbent	-6.63*	-0.49*	-2.81*	0.06
Liberal Candidate	0.06	0.23*	0.27	0.26*
Marginality	-0.23*	0.00	-0.04*	-0.06*
Unemployment	0.30*	0.09*	0.26*	0.03
Working Class O-O	0.01	-0.01	0.01	-0.01
R^2 (adj.)	0.82	0.84	0.79	0.78

voters were much more likely to shift to Labour if the Alliance was represented by a Liberal candidate than if it was represented by a member of the SDP. This suggests that where an SDP candidate was not available, former non-Labour/Liberal voters were much more likely to prefer Labour to Liberal; Labour and SDP were in direct competition. Further, where unemployment was high, former Liberal voters were much more prepared to switch to Labour, suggesting that the latter was still seen as best able to solve the problems of economic depression by a significant number of voters; this was not so, however, among those who did not vote in 1979. The urban:rural dimension so apparent for Conservative trends (Table 7.3) was not present for Labour, however; in general it performed best where it always had.

The regressions for the *Alliance* parties provide relatively weak fits, except in the case of flows from Labour to the Alliance (Table 7.5). These were especially strong in constituencies where the Liberal Party did well in 1979, suggesting very substantial movement away from Labour in seats where the Alliance appeared to have a chance of victory. Liberal and SDP incumbents were particularly successful at winning votes from former Labour supporters. There was, however, a strong regional variability to the shift from Labour to the Alliance. All eight regions incorporated in the equation had, on average, at least two percentage points less than did the Southwest region moving from Labour to the Alliance, and there were significantly smaller shifts in all except East Anglia and Greater London. In other words, it was mainly in the south of England where the Alliance was already strong (via the Liberal Party), unemployment was low and Labour incumbents were few, that this shift occurred in above-average proportions. Former Labour voters were, it seems, especially prepared to switch allegiance to the Alliance where there was an SDP incumbent; the Labour defectors held on to a substantial proportion of the vote that they won when members of that party. However, among the non-incumbents, the Liberal candidates were better able than their SDP counterparts to win-over former Labour votes.

The Alliance attracted 1979 non-voters in very much the same pattern as it did former Labour voters, though the differentials displayed by the regression coefficients were not as great. The major difference between the two equations is that those who did not vote in 1979 were no more attracted by an SDP than by a Liberal incumbent in 1983. Non-voters were no more likely to shift to the Alliance in the safer seats; former Conservative and Labour voters were.

The strength of the regional pattern in the shifts to, and loyalty for, the Alliance does not incorporate all of the geographical sources of variance, however. Maps of the residuals from the four regressions all indicate local concentrations of

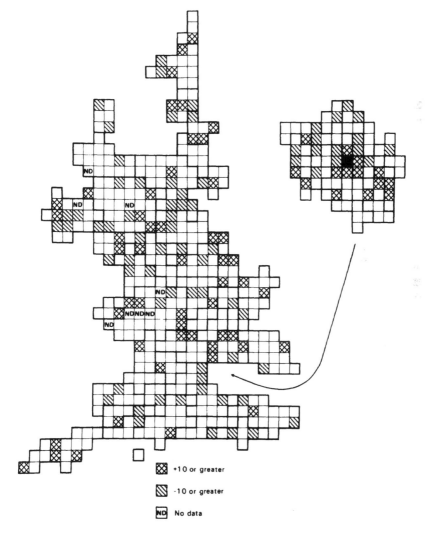

Figure 7.13 Residuals from the Regression on Alliance Loyalty

Table 7.5 Regression Analyses: Flow-of-the-Vote to Alliance

1979 Vote 1983 Vote	Liberal Alliance	Conservative Alliance	Labour Alliance	Did Not Vote Alliance
Constant	70.96	13.71	24.80	12.96
Liberal 1979	0.14*	-0.02	0.34*	0.05*
North	-2.77*	-0.82	-5.68*	-1.49*
Yorks/Humberside	-3.56*	-1.02*	-6.21*	-1.87*
Northwest	-5.13*	-1.41*	-6.73*	-1.96*
West Midlands	-2.37*	-1.17*	-3.66*	-1.01*
East Midlands	-5.11*	-1.90*	-5.78*	-2.20*
East Anglia	-2.23	-1.29*	-2.03	-0.91
Greater London	-6.55	-2.50	-4.85	-2.85*
Southeast	-2.56*	-1.32*	-2.17*	-1.33*
Metropolitan Borough	-0.52	0.50	-1.33	-0.71*
Inner London	-0.52	1.71	-2.94	-1.12
Outer London	1.39	0.99	-0.64	-0.36
Metropolitan County	-1.11	0.13	-2.02*	-0.74
Nonmetropolitan Borough	0.05	0.41	-0.54	-0.37

Conservative Incumbent	0.20	-0.50	0.50	-0.25
Labour Incumbent	-3.22*	-0.73*	-2.97*	-0.89*
Liberal Incumbent	8.31*	7.57*	4.33*	3.22*
SDP Incumbent	7.42*	3.94*	7.31*	3.12*
Liberal Candidate	0.36	0.29	0.42*	0.17
Marginality	0.03	0.02*	0.11*	-0.01
Unemployment	-0.52*	-0.07	-0.53*	-0.19*
Working Class O-O	-0.03	-0.05*	0.02	0.02
R² (adj.)	0.48	0.27	0.71	0.43

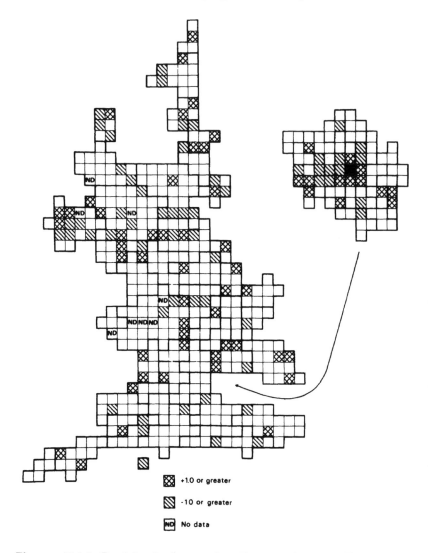

Figure 7.14 Residuals from the Regression on Flows from Conservative to Alliance

Alliance vote-winning ability (and also small blocks of constituencies where the Alliance failed to make substantial inroads). This is shown in Figures 7.13 and 7.14. In the former, relating to Alliance loyalty, there are several clusters of constituencies where the Alliance did much better than predicted: they include parts of Tyneside and Cleveland (where there were several SDP incumbents whose influence apparently spread to adjacent seats);

Crosby (scene of an SDP by-election success in 1982) and neighbouring Bootle; Warrington (where the SDP leader narrowly lost a by-election in 1981 and where the Alliance retained more than a remnant of that support in what was formerly a safe Labour seat); South Humberside and North Lincolnshire; parts of Hertfordshire; parts of Inner London (around the Bermondsey seat won at a by-election by the Liberals in 1983), southeast London (the 'traditional' Liberal stronghold of Orpington) and southwest London (Richmond, Kingston and Sutton); and the city of Plymouth (one of whose constituencies is represented by the current SDP leader, David Owen). The residuals from the regression of Conservative-to-Alliance flows repeat some of these clusters, such as Warrington, Crosby, Teesside and the three London nodes (Figure 7.14): they also reemphasise the poor showing of the Alliance in areas such as the South Yorkshire coalfield, Leicester and inner-west London.

None of the 22 independent variables is significant in all twelve of the equations, indicating that different factors were affecting different elements in the geography of the flow-of-the-vote matrix. The prior strength variable was significant in eleven of the twelve, however, providing strong support to the hypothesis that parties were best able to win votes where they were already strong. (The exception was with the flow of voters from Conservative to the Alliance, a pattern which could not be predicted statistically with much success.)

The eight regional variables together provide substantial evidence of a major north:south division in the pattern of voting change in the country, additional to the prior strength variables. In general terms, relative to the situation in the Southwest, the Conservative Party did less well in the northern regions (especially in winning over those who either voted Labour or did not vote in 1979), as did the Alliance Parties — even with respect to retaining the support of 1979 Liberal voters. The Labour Party, on the other hand, performed much better in the northern than the southern regions — including the Midlands where it successfully retained above-average levels of support, won over substantial numbers of former Liberal voters and non-voters, and lost fewer voters to the other two parties.

Alongside this north:south division of the country was an

urban:rural split, which mainly affected the Conservative Party. This was less able to retain the allegiance of 1979 voters, and to win over voters from other causes, in the metropolitan counties, in the nonmetropolitan boroughs and in Inner London. The Labour Party was better able both to retain support and to win it in the metropolitan boroughs than elsewhere, but the Alliance displayed no general urban:rural division in its vote-winning activities.

Incumbency was important to the three parties which lost the election, but not to the winners. There was no significant regression coefficient relating to Conservative incumbency, indicating that the party in power neither won more votes where it held the seat nor lost them where it did not. The Conservative Party presents itself as the party of the nation as a whole, and on this variable at least its claim is supported. For the other three parties, however, already holding a seat (or having a candidate who had held a nearby seat) was a considerable advantage. All eight coefficients are significant and positive for the two Alliance Parties in Table 7.5, for example, and ten of the other 16 are significant and negative (the expected sign). Clearly, in a losing party a known candidate had an advantage. In the battle between the losing parties, Liberal incumbents were better able to win former Conservative votes, and SDP incumbents to win former Labour votes.

According to the generally-accepted theses regarding tactical voting, loyalty to parties should be greater — and shifts between them less — in the marginal constituencies. This is only partly supported by the regression results. The Conservative and, especially, the Labour Party were both able to retain more support in marginal than in safer seats, but the Alliance was not — largely because in most of the more marginal seats it was not one of the main contestants. Although the flows from Labour to Conservative were less in the more marginal seats, the same did not hold for the reverse flow. And the flows to both Labour and Conservative from the Liberal and non-voters in 1979 increased in the more marginal constituencies. Flows to the Alliance from both Conservative and Labour were greater in the less marginal seats.

Some of the most constant relationships relate to the unemployment variable. The greater the level of unemployment in a constituency, the greater the relative success of the Labour

Party in retaining the support of its 1979 voters, in preventing a flow to either of the other groups, and in attracting defectors. The level of unemployment had no influence on Conservative-Alliance flows, suggesting that this aspect of the government's economic performance was not relevant to the choice between the two.

Whereas one aspect of the government's performance clearly influenced the geography of the flow-of-the-vote, the other did not. There is only one significant regression coefficient for working-class owner-occupation; where this was relatively high, fewer Conservative voters switched to the Alliance.

In Summary

The discussion of local effects in Chapter 4 suggested that, through the long-term processes of voter socialisation, a pattern of party strength in an area is maintained once established. This produces a pattern of sectional effects, whereby continuity in voting behaviour is normal and the national shifts in preferences are damped down. Such sectional effects are clearly illustrated in this analysis of shifts between 1979 and 1983. Each party grouping performed best in the constituencies where it was initially strong, and each grouping was also able to draw on other sectional patterns of support.

The long-lasting sectional effects are occasionally punctuated by local environmental and campaign effects, which later either disappear or generate long-term changes in the voting map. The presence of such effects in 1983 has been illustrated by the analyses here. The level of unemployment in an area clearly influenced partisan choice — whether permanently or not depends on if and when the level falls before future elections. And the arrival of the SDP in 1981 also substantially altered the geography of voting — in 1983 at least. Its incumbents (with one exception defectors from the Labour Party) were apparently able to retain a considerable portion of the votes that they obtained in 1979 as candidates for other parties, and its organisation in constituencies where it lost by-elections in the 1981-3 period also provided it with a substantial increment of support. Finally, its alliance with the Liberal Party suggested a third potential victor to the electorate, and enabled both to garner substantial numbers of votes in parts of the country where the electoral chances of Labour were bleakest.

The national flow-of-the-vote matrix is not reproduced in every constituency, as was indicated in Chapter 2. Instead, there are substantial geographical variations in the rates of party loyalty and inter-party shifting. The analyses here indicate that these variations are probably part of a long-term pattern of sectional effects, invaded somewhat in 1983 by the unemployment consequences of government policies and the development of a new party grouping whose roots were being established in selected locales.

Decomposing the Flow Patterns

It was shown in the previous chapter that there were substantial inter-correlations among the various cells of the estimated matrices. The same is likely to be true with the estimated flow-of-the-vote matrices. A constituency with an above-average Labour loyalty, for example, is likely — all other things being equal — to have above-average flows to Labour from the other parties. (This is, indeed, suggested by the hypotheses.) To some extent, therefore, the analyses just reported may focus mainly on the common patterns and not clarify the differences between the various flows.

That this is likely is shown by the results of three separate principal components analyses, one each for the lists of dependent variables in Tables 7.3-7.5. In each the loadings for all four variables on the principal component are high (Table 7.6), though not as high as those reported in Chapter 6, and so most of the variation is accounted for (86 per cent in the first case, 95 per cent in the second and 82 per cent in the third).

Given the strong inter-correlations shown in Table 7.6, then the detailed appraisal of the different flows in Tables 7.3-7.5 to some considerable extent may have been clouded by the general pattern relevant to a particular party. Because the principal components do not account for all of the variation, the flows have been both generalised and decomposed. They were generalised by taking the scores on the three principal components identified in Table 7.6 to represent the overall pattern for Conservative, Labour and Alliance respectively. They were decomposed by taking each variable in each set of four in turn, regressing it against the other three in that set, and taking the residuals from

Table 7.6 Principal Components Analyses of the Flow-of-the-Vote Matrices

	Conservative	Labour	Liberal	Did Not Vote
A. Conservative Vote, 1983				
1979 Vote				
Loading	0.99	0.85	0.83	0.93
B. Labour Vote, 1983				
1979 Vote				
Loading	0.95	0.98	0.97	0.95
C. Alliance Vote, 1983				
1979 Vote				
Loading	0.69	0.86	0.99	0.95

that regression as indicating the unique component of each variable. (The individual variables were regressed against the other independents rather than against the component scores because the latter were influenced in their values by those of the variable with which they would have been regressed. With only four variables contributing to the component scores, there would probably have been substantial inbuilt correlations, regressing a variable against itself.) Thus, for example, Labour loyalty — in standardised form — was regressed against the Conservative-Labour, Liberal-Labour and Did Not Vote-Labour flows: the residuals from that regression indicated that part of the geography of Labour loyalty that was not common to the other geographies of flows to Labour. Both the component scores and the residuals were than regressed against the standard set of independent variables, the former to evaluate the general geographies of flows and the latter to evaluate the unique elements.

The results of the regressions using the component scores as the dependent variables are given in Table 7.7 (the three maps that these relate to are shown in Figures 7.15-7.17). The strong regional variations stand out. For the Conservative Party, there were only two constituencies in the northern regions (Boothferry in North Humberside and Brigg in South Humberside) where in general the flows were substantially above average, and only Newark, Leominster and Staffordshire South in the Midlands regions were similarly placed. Labour, on the other hand (Figure 7.16) attracted virtually all of its substantially above-average flows in the northern and Midland regions, plus Inner London: the only exceptions were Ipswich and Oxford East (the latter comprising the industrial suburbs of the city). The Alliance, too, shows a clear regional pattern (Figure 7.17). Outside London, there were only two constituencies (Ipswich and Slough) in the southern regions where it recorded a substantially below-average set of flows, and it obtained a component score exceeding +1.0 in all but three constituencies in the Southwest.

These regional variations are identified in the regression analvses (Table 7.7), illustrating the strong Labour showing in the North and Midlands, countered by the strong Alliance showing in the Southwest and Southeast. For the Conservative Party, the only significant regional trend was its below-average showing in the Northwest; for this party, the main spatial divide

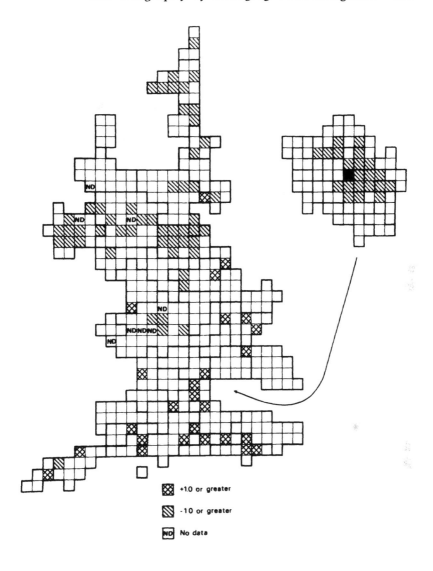

+1.0 or greater

-1.0 or greater

ND No data

Figure 7.15 Scores on the Principal Component for Flows to Conservative

was the urban:rural — it performed best in the nonmetropolitan, county constituencies, and significantly less well nearly everywhere else. Both Conservative and Labour performed better in the marginal seats: the greater the unemployment level, the stronger the flows to Labour and the weaker the flows to the other two.

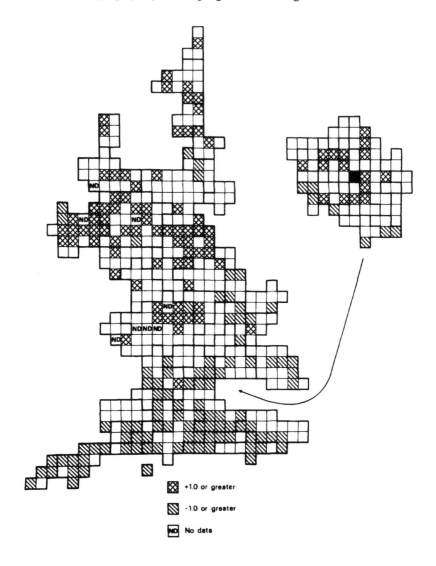

Figure 7.16 Scores on the Principal Component for Flows to Labour

All three parties/groups won most support in their areas of existing strength: the simple correlations between the relevant component scores and the estimated vote in 1979 were — Conservative, 0.74: Labour, 0.86; Alliance, 0.45. The Alliance incumbents attracted substantially above-average flows, and as a consequence reduced the flows to the other two. Among those who voted Alliance, whether it was a Liberal or an SDP

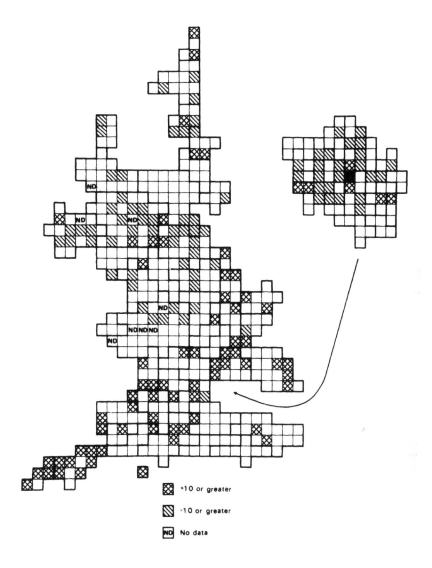

Figure 7.17 Scores on the Principal Component for Flows to Alliance

incumbent made virtually no difference; former Labour voters, however, were more attracted by SDP incumbents and Conservative voters by Liberal incumbents. For the Conservative Party, its incumbent candidates did not attract an above-average flow. Labour incumbents did, however, thereby staunching somewhat the flow to the Alliance.

Table 7.7 Regression Analyses: Principal Components of the Flow-of-the-Vote Matrix

Component	Conservative		Labour		Alliance
Constant	-1.06		-1.65		1.19
	Con.		Lab.		Lib.
1979 Vote	0.05*		0.04*		0.03*
North	-0.16		0.33*		-0.35*
Yorks/Humberside	-0.20		0.28*		-0.49*
Northwest	-0.37*		0.58*		-0.72*
West Midlands	-0.11		0.31*		-0.29*
East Midlands	-0.16		0.33*		-0.74*
East Anglia	-0.06		0.08		-0.31*
Greater London	-0.38		0.12		-1.05*
Southeast	-0.16		0.10		-0.40
Metropolitan Borough	-0.53*		0.21*		-0.14
Inner London	-1.19*		0.33		-0.20
Outer London	-0.32		0.09		0.17
Metropolitan County	-0.26*		0.09		-0.19
Nonmetropolitan Borough	-0.25*		0.01		0.03

Conservative Incumbent	0.01	-0.01	0.02
Labour Incumbent	0.02	0.25*	-0.48*
Liberal Incumbent	-0.78*	-0.33*	1.08*
SDP Incumbent	-0.44*	-0.47*	1.04*
Liberal Candidate	-0.04	0.05	0.05
Marginality	-0.02*	-0.01*	-0.01
Unemployment	-0.05*	0.03*	-0.08*
Working Class O-O	0.01*	0.00	-0.01
R^2 (adj.)	0.74	0.83	0.46

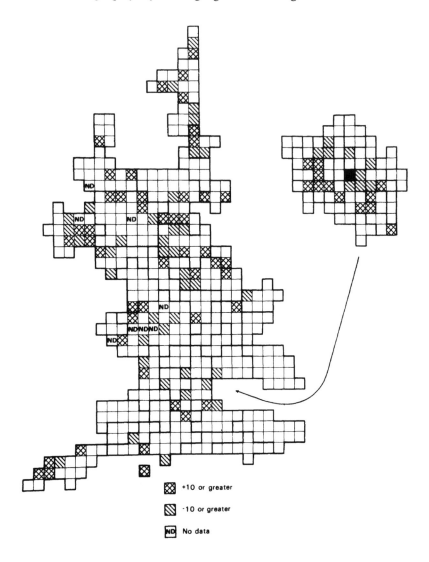

Figure 7.18 Residuals from the Regression on the Principal Component for Flows to Conservative

The regressions reported in Table 7.7 refer to the general patterns of flows. Their residuals indicate the constituencies where the flows were substantially greater or less than those predicted. In general, they suggest small areas of the country where each party was either unexpectedly strong (the positive residuals) or weak (the negatives), and apart from Greater

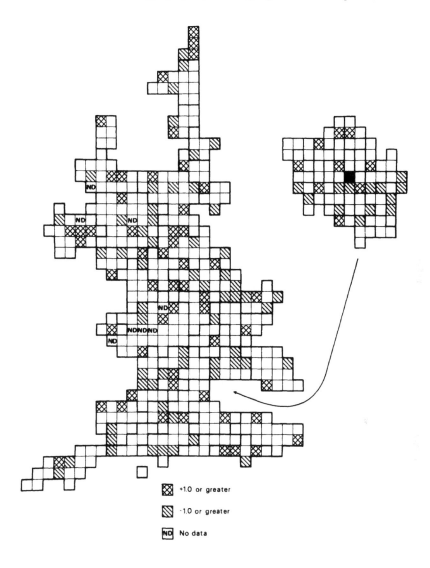

Figure 7.19 Residuals from the Regression on the Principal Component for Flows to Labour

London there are no major spatial concentrations (Figures 7.18-7.20). The Conservative Party attracted a smaller flow than expected in parts of Barnsley and west Sheffield (the more Conservative of the seats there), for example, but a much larger one in Colne Valley and Dewsbury. The Labour Party had unexpected pockets of strength in parts of Oxfordshire and in

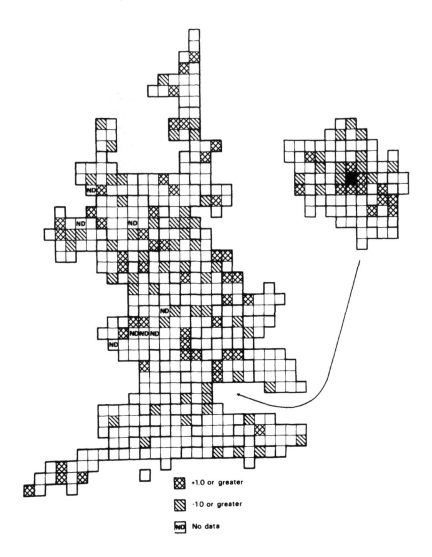

+1.0 or greater

-10 or greater

No data

Figure 7.20 Residuals from the Regression on the Principal Component for Flows to Alliance

Brighton — where it won seats in the 1960s — (Figure 7.19), but performed relatively weakly in the Portsmouth area. The Alliance was weak in several coalfield areas, in Stoke and in Leicester and on the western edge of London (Figure 7.20), but did well in parts of Inner London (reflecting its by-election victory at Bermondsey early in 1983), in Plymouth, in the Grimsby area and in west Coventry. Such nodes of support suggest local environmental effects — probaby reflecting, in part at least, the strength of candidates and organisations.

These general patterns can be removed by regression as indicated above, allowing identification of the constituencies with strengths or weaknesses in particular elements of the flow matrices. Thus, Table 7.8 gives the results for the Conservative Party analyses, for the pattern of flows once the general pattern (analysed in Table 7.7) is excluded. In general, the correlations are weak — only the pattern of Labour-Conservative flows was very predictable, with an emphasis on below-average flows in the northern regions and in the marginal seats. Loyalty to the party was greatest, *ceteris paribus*, where it was strong in 1979 (a result not repeated for the other two parties). It was also relatively strong in the metropolitan constituencies, and in the boroughs in the nonmetropolitan counties, suggesting something of a reactive environmental effect in areas where Conservative supporters are generally in a minority (p. 62). The flows from Liberal to Conservative suggest that Conservative voters were most likely to shift to the new grouping in the regions where the Alliance was relatively weak and Labour strong.

The goodness-of-fit for the regressions of the Labour residuals is also generally poor (Table 7.9). The first two columns give a clear indication of regional variations, however: Labour loyalty was relatively high in the north, and the shift from Conservative to Labour was relatively small there. Conservative voters in 1979 did shift to Labour in above-average proportions in the seats with Alliance incumbents, however, and also, unsurprisingly, where unemployment was high. Interestingly, those residual patterns show only one (and then a negative) relationship with prior Labour strength.

Turning to the regressions for the Alliance (Table 7.10), in general the correlations are relatively large. They suggest that, once the general pattern has been removed, there were specific geographies to each flow, which differed substantially. Alliance loyalty was relatively high in the northern regions, but relatively few voters were attracted from those who supported Labour there in 1979. Votes were picked up from former Conservative supporters in the urban constituencies, and where unemployment was high but marginality low — suggesting the movement of protesters away from the government party where it was unlikely to influence the local result.

Table 7.8 Regression Analyses: Flows to Conservative with General Pattern Removed

1979 Vote 1983 Vote	Conservative Conservative	Labour Conservative	Liberal Conservative	Did Not Vote Conservative
Constant	-4.34	-3.31	0.12	5.24
Conservative 1979	0.07*	0.03*	-0.02*	-0.06*
North	0.77*	-0.60*	0.38*	-0.59*
Yorks/Humberside	0.44	-0.29	0.65*	-0.56*
Northwest	0.06	-0.64*	0.92*	-0.19
West Midlands	0.02	-0.74*	0.66*	0.13
East Midlands	0.36	-0.44*	1.00*	-0.50*
East Anglia	-0.04	-0.22	0.63	-0.07
Greater London	0.25	0.48	1.25	-1.11
Southeast	0.20	0.02	0.49*	-0.44*
Metropolitan Borough	0.64*	0.26	-0.52*	-0.85*
Inner London	-0.50	0.43	0.14	-0.47
Outer London	1.22	0.11	1.03	-1.04
Metropolitan County	1.07*	-0.07	-0.53*	-0.91*
Nonmetropolitan Borough	0.66*	0.16	-0.32*	-0.72*

Conservative Incumbent	-0.04	0.06	0.12	-0.03
Labour Incumbent	0.44*	-0.26*	0.31*	-0.35*
Liberal Incumbent	-1.79*	1.25*	-0.55	0.72
SDP Incumbent	-1.11*	0.49*	-0.71*	0.75*
Liberal Candidate	-0.21	0.08	-0.04	0.14
Marginality	-0.03*	0.07*	0.01	-0.02*
Unemployment	0.05	0.02	0.03*	-0.09
Working Class O-O	0.03*	0.01	0.01	-0.02
R² (adj.)	0.32	0.65	0.18	0.27

Table 7.9 Regression Analyses: Flows to Labour with General Pattern Removed

1979 Vote 1983 Vote	Labour Labour	Conservative Labour	Liberal Labour	Did Not Vote Labour
Constant	−1.90	−1.35	−0.18	1.62
Labour 1979	0.02	0.01	−0.02*	−0.01
North	2.29*	−0.42*	−0.53*	−0.13
Yorks/Humberside	2.62*	−0.23*	−0.44	−0.45*
Northwest	1.42*	−0.21*	0.18	−0.13
West Midlands	1.93*	−0.42*	−0.28	−0.09
East Midlands	2.63*	−0.36*	0.21	−0.49*
East Anglia	1.80*	−0.23	−0.02	−0.37
Greater London	2.74	−0.08	0.67	−0.98
Southeast	1.36*	−0.11	0.18	−0.36*
Metropolitan Borough	−0.25	0.32*	0.09	−0.16
Inner London	1.63	0.52	−0.59	−0.73
Outer London	0.75	0.13	−0.42	−0.22
Metropolitan County	0.70	−0.04	0.15	−0.20
Nonmetropolitan Borough	0.52	0.01	−0.16	−0.13

	(1)	(2)	(3)	(4)
Conservative Incumbent	1.41*	-0.11	-0.33*	-0.28*
Labour Incumbent	-1.07*	-0.07	0.86*	0.27*
Liberal Incumbent	-2.04*	0.90*	-1.57*	0.23
SDP Incumbent	-2.21*	0.45*	-0.87*	0.34
Liberal Candidate	-1.04*	0.15*	0.10	0.20*
Marginality	-0.06*	0.03*	0.01	-0.02*
Unemployment	-0.03	0.04*	0.12*	-0.05*
Working Class O-O	0.03	-0.01*	0.01	-0.01
R^2 (adj.)	0.28	0.56	0.21	0.27

Table 7.10 Regression Analyses: Flows to Alliance with General Pattern Removed

1979 Vote 1983 Vote	Liberal Alliance	Conservative Alliance	Labour Alliance	Did Not Vote Alliance
Constant	1.92	-1.67	-8.74	2.27
Liberal 1979	-0.08*	-0.03*	0.17*	-0.01
North	1.93*	0.07	-1.73*	-0.05
Yorks/Humberside	1.83*	0.23	-1.22*	-0.24
Northwest	0.79*	-0.08	-1.46*	-0.06
West Midlands	1.01*	-0.48	-1.17*	0.12
East Midlands	0.72	-0.22	-0.17	-0.30
East Anglia	0.21	-0.54	-0.01*	-0.01
Greater London	-0.41	-0.01	2.35	-0.87
Southeast	0.33	-0.16	1.02*	-0.38*
Metropolitan Borough	0.57	1.07*	0.98*	-0.63*
Inner London	1.27	2.51*	1.26	-1.19*
Outer London	1.60	1.23	0.69	-0.63
Metropolitan County	0.56	0.67*	0.24	-0.38
Nonmetropolitan Borough	0.48*	0.72*	0.72*	-0.43*

Conservative Incumbent	0.36	-0.19	0.82*	-0.17
Labour Incumbent	-0.55*	-0.10	-0.49	0.02
Liberal Incumbent	-0.09	4.59*	-0.96	-0.57
SDP Incumbent	-0.92*	1.45*	0.12	0.13
Liberal Candidate	-0.14	0.16	0.09	-0.03
Marginality	-0.03*	0.05*	0.15*	-0.04*
Unemployment	-0.05*	0.08*	0.04	-0.06*
Working Class O-O	0.01	-0.03*	0.05*	-0.01
R² (adj.)	0.23	0.50	0.61	0.45

And Those Who Did Not Vote?

So far, no analyses have been reported for non-voters in 1983, neither the 'loyal abstainers' (those who did not vote in 1979 as well) nor those who voted in 1979 but not in 1983. No hypotheses were presented regarding these types of voters in Chapter 4. Nevertheless, there were considerable inter-constituency variations in the size of these groups (Table 7.2), some of which may represent the operation of local effects. In particular, one environmental effect could be anticipated:

> The more marginal a constituency, the smaller the percentage of 'loyal abstainers' and the smaller the flow of people who voted in 1979 to the non-voting category in 1983.

This hypothesis is tested here. Other independent variables are included in the regressions (the same set as previously used in this chapter) to inquire whether there were regional and other variations. Further, a 'safe-seat' hypothesis was tested, suggesting that a party is more likely to be deserted, by voters who do not vote in the second contest, when votes for the party are irrelevant, because the chances of victory are remote:

> The greater the percentage of the vote won by a party in 1979, the smaller the percentage of its supporters then who did not vote in 1983.

The results of the four regressions are presented in Table 7.11. The first hypothesis is in line with all four: the safer the seat, the greater the percentage of 1979 non-voters who failed to vote again in 1983 and the greater the flow to the non-voter category from support for the three parties in 1979. The second hypothesis is consistent with the results in only two of the regressions; the Conservative and the Labour Parties both lost fewer votes to the abstentions category in the constituencies where they were strong than in those where they were weak. (So too did the Liberal Party, but its negative regression coefficient was statistically insignificant. Note, however, substantially smaller losses in constituencies where the Alliance parties fielded incumbent candidates in 1983.)

Turning to the other variables in the equations, clear regional and urban:rural differences are indicated. Both the Conservative and the Liberal Parties lost substantially more votes to the non-voting category in several of the northern regions than they did in the Southeast — regions where in general those parties were relatively weak from the outset — whereas the Labour Party did not. The flow to non-voting was significantly above the Southwest level in all four categories in the Southeast and Greater London regions. In addition, both the shift to non-voting in 1983 (from whatever party) and remaining as a non-voter between 1979 and 1983 were in general significantly greater in the metropolitan areas, and in the nonmetropolitan boroughs, than they were in the shire counties.

A final, salient variable in the four equations is that for unemployment. In all four, the greater the 1981 level of unemployment the greater the flow into, and remaining in, the non-voting category. This can be interpreted as a 'political alienation' effect. The greater the local economic crisis, the more disillusioned residents were with the ability of any party to produce a solution. The urban:rural and regional significant relationships can be linked with this, in that the urban and regional environments provide a wider context within which local experience of the unemployment problem can be set.

As with the earlier analyses in this chapter, to some extent the pattern is reproduced in each equation, reflecting a general trend with regard to the geography of non-voting in 1983. Thus the same decomposition procedure was undertaken, involving a principal components analysis, a regression analysis of each flow against the other three, and separate regressions for the residuals.

The principal component accounted for 73 per cent of the variation in the four flow sets, with loadings as follows:

1979	1983	
Conservative	Did Not Vote	0.89
Labour	Did Not Vote	0.48
Liberal	Did Not Vote	0.90
Did Not Vote	Did Not Vote	0.99

The scores are shown in Figure 7.21. By far the dominant feature of this map is the concentration of positive scores, indicating above-average flows in all four categories, in Greater London,

Table 7.11 Regression Analyses: Flows to the Did Not Vote Category

1979 Vote 1983 Vote	Did Not Vote Did Not Vote	Conservative Did Not Vote	Labour Did Not Vote	Liberal Did Not Vote
Constant	56.44	10.37	13.33	5.80
Vote 1979	—	Con. −0.10*	Lab. −0.10*	Lib. −0.03
North	1.64	0.76	0.23	0.89
Yorks/Humberside	2.51*	1.32*	0.77	1.49*
Northwest	1.80*	1.51*	−0.10	1.62*
West Midlands	0.24	0.35	−0.34	0.40
East Midlands	2.50*	1.27*	0.67	1.75*
East Anglia	0.93	0.60	0.54	0.80
Greater London	5.47*	3.13*	3.19*	3.67*
Southeast	2.13*	1.21*	1.12*	1.21*
Metropolitan Borough	3.07*	1.93*	0.60	0.95
Inner London	5.46*	4.20*	1.24	2.02*
Outer London	2.73	1.15	0.47	0.15
Metropolitan County	2.61*	1.42*	0.79	1.13*
Nonmetropolitan Borough	2.13*	1.02*	1.00*	0.66*

Conservative Incumbent	0.45	0.07	0.36	0.05
Labour Incumbent	0.42	0.12	-0.28	0.64
Liberal Incumbent	-0.17	-0.05	-0.82	-1.83*
SDP Incumbent	-1.62	-0.32	-0.42	-2.06*
Marginality	0.13*	0.07*	0.11*	0.05*
Unemployment	0.42*	0.28*	0.09*	0.26*
Working Class O-O	0.02	-0.01	0.02	0.01
R^2 (adj.)	0.46	0.67	0.57	0.45

Table 7.12 Regression Analyses of Non-voting: Principal Component and Flows with General Pattern Removed

Vote 1979 / Vote 1983	Principal Component	Did Not Vote Did Not Vote	Conservative Did Not Vote	Labour Did Not Vote	Liberal Did Not Vote
Constant	-2.19	-0.61	2.74	0.35	-0.53
Vote 1979	—	—	Con. -0.05*	Lab. -0.03*	Lib. -0.02
North	0.36*	0.55*	-0.63*	-0.90*	0.18
Yorks/Humberside	0.53*	0.32	-0.48*	-0.49	0.34
Northwest	0.40*	-0.02	-0.15	-0.36	0.89*
West Midlands	0.07	0.15	-0.23	-0.51	0.43
East Midlands	0.53*	0.41	-0.65	-0.75*	0.70*
East Anglia	0.20	0.01	-0.16	-0.09	0.50
Greater London	1.11*	-0.30	-0.10	0.71	1.28
Southeast	0.43*	0.20	-0.21	-0.16	0.30
Metropolitan Borough	0.62*	0.82*	-0.27	-0.89*	-0.61*
Inner London	1.10*	0.38	0.53	-0.05	-0.81
Outer London	0.53	1.47	-0.74	-1.61	-1.35
Metropolitan County	0.54*	0.51*	-0.39	-0.57	-0.14
Nonmetropolitan Borough	0.41*	0.51*	-0.24	-0.37	-0.41*

Conservative Incumbent	0.07	0.36*	-0.13	-0.31	-0.16
Labour Incumbent	0.12	0.04	-0.41*	-0.38*	0.52*
Liberal Incumbent	-0.09	0.42	0.61*	0.12	-2.43*
SDP Incumbent	-0.35*	-0.13	0.80*	0.53	-1.47*
Marginality	0.02*	-0.02*	0.03*	0.06*	-0.02*
Unemployment	0.09*	-0.01	-0.02	-0.01	0.07*
Working Class O-O	0.03	0.02	-0.01	-0.02	0.01
R² (adj.)	0.48	0.13	0.34	0.41	0.29

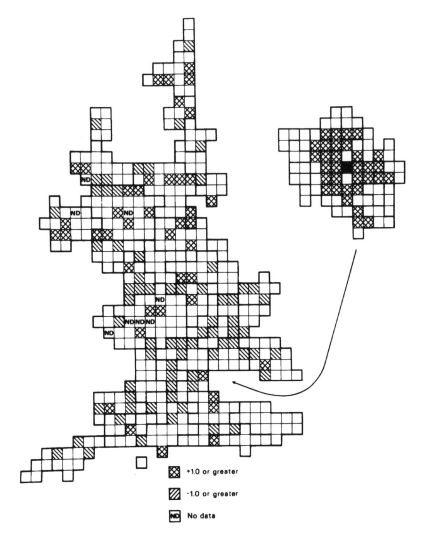

Figure 7.21 Scores on the Principal Component for Flows to Non-voting

especially, but not entirely, Inner London. Countering this, much of the northern Home Counties and East Anglia have large negative scores, indicating relatively small flows into the non-voting category; there are also concentrations of such scores in parts of Gloucestershire, the north West Midlands and northwest Lancashire. The regression analysis (the first column of Table 7.12) picks out the correlates of this pattern; in general, the non-voting 'destination' was favoured most in the northern regions and in Greater London, in the metropolitan, borough and Inner London constituencies, in the safer seats, and in the places with high levels of unemployment.

Turning to the individual flows from which this general component has been removed, the other four regressions show that for Conservative and Labour the two initial hypotheses account for much of the particular variation: each party lost fewer 1979 supporters in the more marginal seats and in those where it was strong. Both also lost fewer in constituencies with Labour incumbents; the Conservative Party lost more where Alliance candidates were standing. The pattern for those who did not vote at both elections differed from the general in the urban:rural split and in the smaller percentages involved in the marginal constituencies. Those who voted Liberal in 1979 were more likely to abstain in 1983, relative to other 1979 voters, in areas with high unemployment and in constituencies with Labour incumbents, but less likely to abstain in the metropolitan areas and in constituencies where its incumbents were candidates.

In Summary

The analyses in this chapter have provided strong circumstantial evidence in favour of the various hypotheses outlined in Chapter 4. The estimating procedure suggests substantial variations among constituencies in the relative size of the 16 cells in the flow-of-the-vote matrix; those variations are especially large for flows involving Labour voters. Regression analyses using the estimated values as dependent variables have been extremely successful in accounting for much of this variation, in terms of generally-accepted models of voting behaviour. In brief, these indicate patterns entirely consistent with the sectional effect — the parties performed best in 1983, in relative terms, where they did best in 1979. The nature of the local context (its marginality and the candidates) had some influence on vote-switching, as did the local environment (notably the level of unemployment). Overall, continuity in the electoral geography of England was retained, with some local variations. The full implications of this are explored in the final chapter.

8   CLASS AND CHANGING VOTES

The two previous chapters have demonstrated that both the geography of voting in England in 1983 and the geography of changing party preferences there between 1979 and 1983 are highly predictable. Using a combination of independent variables representing the socio-economic characteristics of constituencies, their regional location and the electoral context, it has been possible to account for up to 85 per cent of the inter-constituency variation. Many of the major residuals from the regression models can be accounted for in terms of either particular local partisan loyalties (e.g. the 'Sheffield effect' in South Yorkshire/ North Derbyshire, reflected also in the nature and policies of many of the local governments there) or local circumstances pertaining to the 1983 election.

The general interpretation of these findings (to be developed more fully in the concluding chapter) is of continuities in voting patterns, whereby particular classes have been socialised to support certain parties, more strongly in some regions than in others. The growth of the Alliance did not alter these continuities very substantially — except in a few particular places. This suggests that it should be possible to cast further light on the 1983 result by combining the cross-sectional approach of Chapter 6 with the geography of change approach of Chapter 7. This is done here with an analysis of the flow-of-the-vote matrix for each of the two main social classes.

Class and the Flow-of-the-Vote

Using the BBC/Gallup survey data file, a separate flow-of-the-vote matrix was established for middle- and working-class respondents. As with the other matrices established in this way, there was an underestimation of the numbers not voting and a slight overestimation of the support given to the three parties. To remove these, the matrix was smoothed in the usual way (see

Chapter 5 and Appendix 3) to produce a national 'flow-of-the-vote by class' (8 × 4) matrix whose row and column sums conformed closely to the national 1979 and 1983 distributions of votes. (There are two columns for each party — the national 1983 vote for Conservative in each class, for example; these were summed for comparison with the 1983 results.) The resultant matrix is presented in Table 8.1, broken into two parts and with the cell values expressed as percentages of the row total for *the relevant class*.

Table 8.1 The Flow-of-the Vote by Social Class

A. Middle Class

1979	Conservative	Labour	1983 Alliance	Did Not Vote
Conservative	76.1	2.2	10.0	11.8
Labour	5.7	57.0	18.9	18.4
Liberal	14.7	3.9	64.3	17.1
Did Not Vote	38.0	17.8	16.4	32.8

B. Working Class

1979			1983	
Conservative	65.9	3.1	10.9	20.1
Labour	5.1	57.3	15.9	21.6
Liberal	7.2	6.6	65.9	20.4
Did Not Vote	16.1	17.9	17.4	48.6

Inspection of the table indicates considerable differences between classes in some cells, but similarities in others. For Labour voters in 1979, for example, there is little variability between classes in their 1983 destinations — slightly fewer working- than middle-class Labour supporters shifted to the Alliance, and more to the non-voting category. Among 1979 Conservative voters, however, there was much greater 1983 loyalty among the middle class; nearly twice the percentage of working-class Tories in 1979 shifted to non-voting in 1983, indicative of greater working-class condemnation of the record of the Thatcher government.

Of those who voted Liberal in 1979, there was little variation between the classes in the percentage remaining loyal to the Alliance. Twice as many middle-class as working-class 1979 Liberals switched to Conservative in 1979, however, with more working-class voters switching to Labour. Finally, among those who did not vote in 1979, many more in the working class than in the middle class remained in that category in 1983. Of those who voted at the second contest, in the middle class more switched to Conservative than to the other two parties combined; in the working class, the distribution among the three was approximately equal.

Given this variation between the classes, the question that arises — following the analyses of the previous two chapters — is whether the percentages varied substantially among the 515 constituencies (excluding the eight with no Liberal presence in 1979: p. 164). To test this, the entropy-maximising estimation procedure was used once more. This involved two stages. First, the vote by class (middle and working) in 1983 was estimated for each constituency. These estimates (eight per constituency: two classes and four voting categories) were used as the column constraints in the second stage; the row constraints were provided by the 1979 (BBC/ITN estimates) voting by constituency and the matrix constraint was the national flow-by-class matrix (presented in percentage form in Table 8.1). After only four iterations this provided a satisfactory set of estimates; the cell values were expressed as percentages of the relevant class-row totals, as in Table 8.1.

The Parameters of the Geography

Frequency distributions for the 32 cell values are set out in Figures 8.1-8.4, with the parameters of those distributions in Tables 8.2 and 8.3. As with all of the other estimated values investigated in this book, they indicate very substantial geographical variability in English voting behaviour.

For both classes, in relative terms the greatest variability is in the percentages who voted Labour in 1983 but not in 1979, with four of the six coefficients of variation exceeding 0.5. The percentages involved are very small, and very much positively skewed, with regard to 1979 Conservative and Liberal voters (Figures 8.1, 8.3); for 1979 non-voters, the percentages are larger, and the frequency distributions more platykurtic (Figure 8.4). What these distributions clearly suggest is that the ability of Labour to win over supporters from other causes varied very substantially, by both class *and* place.

The frequency distributions involving 1979 Labour voters also in general have high variability, except with regard to the flow

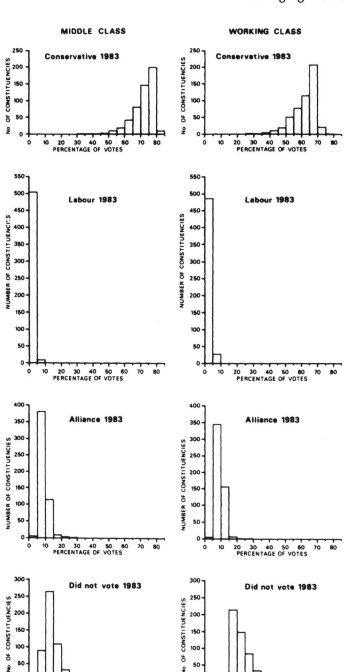

Figure 8.1 Frequency Distributions for Flows from Conservative, 1979

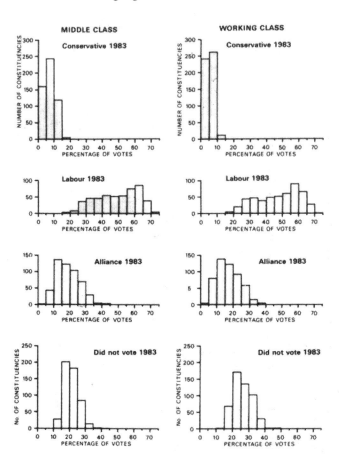

Figure 8.2 Frequency Distributions for Flows from Labour, 1979

from 1979 Labour voting to 1983 non-voting. Thus, for example, Labour loyalty is much more geographically variable in both classes than is either Conservative or Liberal-Alliance loyalty; the Labour loyalty frequency distributions (Figure 8.2) are close to rectangular, with a slight negative skew, whereas the other two are much more peaked (Figures 8.1, 8.3). Again this suggests substantial inter-constituency variation in the Labour Party's fortunes, this time in its ability to retain the support of its 1979 voters. This is matched by the variability in the flows away from Labour.

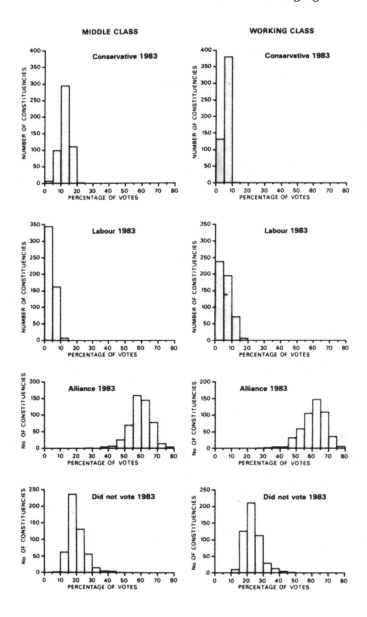

Figure 8.3 Frequency Distributions for Flows from Liberal, 1979

Clearly, then, there is a geography of changing support by class to be accounted for. The remainder of this chapter essays such an account, in terms of the models already tested.

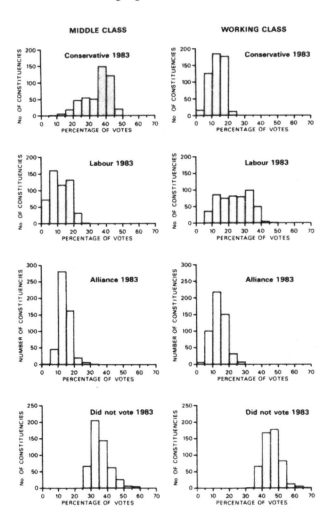

Figure 8.4 Frequency Distributions for Flows from Non-voting, 1979

Regression Analyses

The regression models used here incorporate all of the independent variables used in the previous two chapters. Thus it is anticipated that constituency socio-economic and electoral characteristics, plus location, will provide excellent predictors of the flow-of-the-vote within each class. (The socio-economic

Table 8.2 The Constituency Flow-of-the-Vote Matrices: Summary Statistics for the Middle Class

1979	1983	National Figure	Mean	Standard Deviation	Coefficient of Variation	Minimum	Maximum
Conservative	Conservative	76.1	72.6	6.9	0.10	33	83
	Labour	2.2	2.5	1.4	0.56	0	7
	Alliance	10.0	10.0	2.4	0.24	3	32
	Non-vote	11.8	14.9	4.9	0.33	8	37
Labour	Conservative	5.7	8.3	3.3	0.40	1	16
	Labour	57.0	49.9	12.9	0.26	17	75
	Alliance	18.9	19.7	7.1	0.36	4	43
	Non-vote	18.4	22.1	4.5	0.20	12	42
Liberal	Conservative	14.7	13.7	3.0	0.22	2	13
	Labour	3.9	4.8	2.5	0.52	0	13
	Alliance	64.3	60.4	6.5	0.11	29	80
	Non-vote	17.1	21.1	5.2	0.25	10	43
Non-vote	Conservative	38.0	35.4	7.8	0.22	9	51
	Labour	17.8	12.5	5.5	0.44	2	26
	Alliance	16.4	15.3	3.5	0.23	4	32
	Non-vote	32.8	36.8	5.8	0.16	26	62

Table 8.3 The Constituency Flow-of-the-Vote Matrices: Summary Statistics for the Working Class

1979	1983	National Figure	Mean	Standard Deviation	Coefficient of Variation	Minimum	Maximum
Conservative	Conservative	65.9	62.9	7.6	0.12	26	77
	Labour	3.1	3.0	1.7	0.57	0	8
	Alliance	10.9	10.4	2.3	0.22	3	28
	Non-vote	20.1	23.7	6.2	0.26	13	50
Labour	Conservative	5.1	6.3	2.5	0.40	1	12
	Labour	57.3	48.3	13.1	0.27	16	74
	Alliance	15.9	19.7	7.1	0.36	4	40
	Non-vote	21.6	27.2	5.7	0.21	14	49
Liberal	Conservative	7.2	6.9	1.5	0.22	1	11
	Labour	6.6	7.0	3.6	0.51	1	18
	Alliance	65.9	61.6	7.7	0.13	20	80
	Non-vote	20.4	24.4	5.4	0.22	12	47
Non-vote	Conservative	16.6	14.1	4.2	0.30	3	24
	Labour	17.9	24.4	9.1	0.37	6	47
	Alliance	17.4	14.7	4.3	0.29	3	27
	Non-vote	48.6	46.8	5.1	0.11	34	68

variables were not included in Chapter 7 but are here, because of their importance to the results in Chapter 6.)

The first set of eight equations, for the flows terminating with a *Conservative* vote in 1983, fully justify this anticipation; only in the cases of flows from Liberal voting in 1979 to Conservative voting in 1983 are the adjusted R^2 values less than 0.80 (Table 8.4).

Only three of the independent variables are statistically significant in all eight equations. The Conservative Party was better able to retain the support of its 1979 voters in both social classes, and to woo those who voted in other ways then, in constituencies where it did well in 1979, which had substantial agricultural workforces, and which had low levels of unemployment. Between 1979 and 1983, the party was most successful in the relatively prosperous, more rural parts of England, where it has always done well in recent decades. In relative terms, the unemployment variable had most impact. Among the middle classes, for example, Conservative loyalty fell by 0.55 percentage points for every one percentage point increase in the 1981 unemployment level, and its ability to attract 1979 non-voters fell by 0.74 for the same increase.

Apart from these three influences, in most aspects of the flow to Conservative the party was less successful in the constituencies with substantial numbers of workers in energy industries (miners stood out against the general trends in the country, it seems), in the metropolitan counties and Inner London, and in seats with Alliance incumbents. The flows of 1979 Liberals towards the Conservative Party were not affected by these variables (except those referring to Alliance incumbents), however. In general, too, the flows to Conservative were less in the safer seats; the main exceptions to this were with flows from Labour to Conservative for which, in both classes, there was a significant relationship with seat safety. Voters switched from Labour to Conservative where it seemed unlikely to influence the result, it seems, whereas 1979 Conservatives remained loyal where it appeared that their votes mattered, and 1979 Liberals were similarly more likely to switch to the defending government in the marginal contests.

There was a significant urban:rural dimension to the flows, as already noted with regard to the significant coefficients for the metropolitan borough, Inner London and nonmetropolitan

Table 8.4 Regression Analyses: Flows to Conservative, by Class

1979 Vote	Middle Class				Working Class			
	Conservative	Labour	Liberal	Did Not Vote	Conservative	Labour	Liberal	Did Not Vote
Constant	67.88	6.83	8.32	33.87	61.48	5.59	4.72	14.67
Conservative 1979	0.30*	0.10*	0.15*	0.26*	0.28*	0.07*	0.07*	0.11*
Agriculture	0.17*	0.11*	0.09*	0.22*	0.17*	0.08*	0.04	0.12*
Energy	-0.17*	-0.13	-0.03	-0.26*	-0.21*	-0.10*	-0.02	-0.16*
Manufacturing	-0.02	-0.02	0.01	-0.01	-0.07	-0.02	0.00	-0.01
Unemployment	-0.55*	-0.26*	-0.14*	-0.74*	-0.72*	-0.20*	-0.09*	-0.40*
Working Class O-O	0.04	-0.01	0.00	0.00	0.02	-0.01	-0.00	-0.01
Marginality	-0.11*	0.05*	-0.03*	-0.04*	-0.13*	0.04*	-0.02*	0.00
North	0.61	-0.94*	0.73	-0.69	0.53	-0.71*	0.41	-0.86*
Yorks/Humberside	-0.68	-0.83*	0.68	-1.49*	-0.95	-0.59*	0.40	-0.99*
Northwest	-0.55	-1.16*	0.86*	-1.56*	-0.91	-0.87*	0.41	-1.22*
West Midlands	0.08	-0.50	0.65	-0.51	0.04	-0.33	0.39	-0.43
East Midlands	0.26	-0.48	1.11*	-0.57	-0.07	-0.36	0.58*	-0.56
East Anglia	-1.04	-0.83*	0.25	-1.50	-1.04	-0.55	0.21	-0.80
Greater London	-2.45	-0.82	0.94	-3.28	-2.80	-0.57	0.61	-1.52
Southeast	-0.71	-0.24	0.47	-1.16*	-0.88	-0.14	0.30	-0.51

Metropolitan Borough	−2.31*	−0.98*	−1.14	−3.25*	−3.11*	−0.81*	−0.69*	−1.84*
Inner London	−6.24*	−2.19*	−2.04	−6.27*	−7.48*	−1.78*	−1.23	−3.53*
Outer London	−0.91	−0.77	−1.09	−1.73	−1.64	−0.69	−0.67	−1.29
Metropolitan County	−0.03	−0.46	−0.49	−1.04	−0.48	−0.41	−0.30	−0.81
Nonmetropolitan Borough	−1.16*	−0.58*	−0.50	−1.84*	−1.64*	−0.50*	−0.31*	−1.16*
Conservative Incumbent	0.19	0.02	0.21	0.40	0.23	0.01	0.11	0.16
Labour Incumbent	0.19	−0.53*	0.51	−0.93*	−0.10	−0.43*	0.23	−0.77*
Liberal Incumbent	−3.78*	−0.14	−2.65*	−1.89	−4.20*	−0.24	−1.52*	−0.94
SDP Incumbent	−3.64*	−0.42	−2.45*	−2.54*	−3.35*	−0.30	−1.23*	−0.87*
Liberal Candidate	−0.09	0.16	−0.09	0.14	−0.12	0.10	−0.08	0.11
R^2 (adj.)	0.81	0.88	0.64	0.85	0.81	0.87	0.63	0.85

borough constituency types. But there were few significant regional variations. The Conservatives won over fewer former Labour voters and former non-voters in the three northern regions, and also where a Labour incumbent was standing, indicating an inability — in relative terms at least — to invade the Labour 'heartlands'. The flow of 1979 Liberals to the Conservatives was significantly greater in the East Midlands, a region where the Alliance generally performed badly. Finally, it is worth noting that Conservative incumbents won no additional support: incumbent candidates of the party of government picked up no 'personal vote', it seems, unlike those of the opposition parties.

The flow of votes to *Labour* was also highly predictable, according to the results of the regressions (Table 8.5), with no adjusted R^2 value below 0.82. The only entirely consistent influences were Labour incumbency, the unemployment levels and the percentage employed in energy; sitting Labour members obtained substantial advantages over non-incumbents, and all Labour candidates benefited from high local levels of unemployment and from a substantial presence of miners and power-workers. In seven of the eight regressions, too, there were significant relationships with both the Labour vote in 1979 — Labour held on to more voters, and won more over, where it was already strong — and marginality — Labour was less successful in the safer seats. Where Alliance incumbents were standing, Labour was less able to hold on to its 1979 supporters (especially so if the Alliance incumbent were a member of the SDP, suggesting that the Labour 'defectors' took a substantial body of voters with them) and also to win over former 1979 Liberals (especially where the Alliance incumbent was a member of the Liberal Party). Labour was better able to win over former Conservative voters in both classes and former non-voters in the working class where the Alliance candidate was a Liberal.

Relatively few of the coefficients for the regional and constituency type variables are significant in Table 8.5. The Labour Party retained much of its support, in both classes, in the northern and Midlands regions; it won over former Conservatives, Liberals and non-voters most substantially in the Northwest, and also won over former Liberals in the East Midlands. Regarding the constituency type classifications, the only significant coefficients show relatively greater success for Labour in winning the support of those who did not vote for it in 1979 in the inner areas of the metropolitan counties, excluding London. It did not do significantly better-than-average at retaining the support of 1979 Labour voters in those areas, however; all of the metropolitan counties are in the north and Midlands regions, so Labour did no better in the inner cities than elsewhere in those regions, *ceteris paribus*.

The results in Tables 8.4 and 8.5 show both very predictable patterns of flows and very substantial similarities between the middle and working classes. The latter but not the former is true for the flows involving the *Alliance* (Table 8.6), where the average R^2 value is only 0.51.

Two sets of variables dominate this table — those relating to region and to incumbency. Of the 64 regional regression coefficients, all are negative and only 13 are statistically insignificant. In all regions both the flows to Alliance and the retention of 1979 Liberal support was less than in the Southwest; that most of the coefficients are significant indicates substantial intra-regional similarity in the trends. In general, voters in the northern regions and in Greater London were less likely to shift to the Alliance. So were voters in seats where a Labour incumbent was standing — suggesting that the Alliance was best able to compete with the established opposition where the competitor was relative unknown. Alliance incumbents were much more successful than non-incumbents: SDP incumbents were better at winning over former Labour voters, whereas those from the Liberal Party were better at retaining Liberal support and at winning over former Conservatives and non-voters.

Apart from these two sets of variables, the only other to have a general impact was unemployment. The greater the level of unemployment in 1981 the less able were Alliance candidates to win support; in relative terms (i.e. comparing regression coefficients to constant terms) they were weakest at winning over former Labour voters and non-voters, suggesting that the Alliance was not viewed as a credible alternative by those living in the areas of greatest economic distress. The Alliance candidates did do slightly better, especially at winning over former Labour voters, where the Liberal Party was strong in 1979 — suggesting a continuing process of Alliance replacing Labour in some areas; the outflow of Conservative voters was not related to Liberal strength however. The Alliance was also better able to woo former Labour voters in the safer seats, but not in the

Table 8.5 Regression Analyses: Flows to Labour, by Class

1979 Vote	Middle Class				Working Class			
	Conservative	Labour	Liberal	Did Not Vote	Conservative	Labour	Liberal	Did Not Vote
Constant	-0.80	30.99	0.35	2.97	-0.58	29.61	0.62	11.54
Labour 1979	0.05*	0.41*	0.05*	0.16	0.06*	0.45*	0.09*	0.29*
Agriculture	0.00	-0.13	0.01	-0.01	0.00	-0.13	0.01	-0.06
Energy	0.06*	0.34*	0.10*	0.21*	0.07*	0.33*	0.15*	0.31*
Manufacturing	0.01*	0.05	0.02	0.04*	0.01*	0.06	0.03	0.07*
Unemployment	0.09*	0.41*	0.17*	0.25*	0.09*	0.28*	0.21*	0.20*
Working Class O-O	-0.00	0.04	0.01	0.01	-0.00	0.03	0.01	0.02
Marginality	-0.00	-0.29*	-0.03*	-0.08*	-0.01*	-0.29*	-0.05*	-0.19*
North	0.04	4.85*	0.35	1.23*	0.16	4.82*	0.63	2.85*
Yorks/Humberside	0.08	3.44*	0.27	0.58	0.08	2.97*	0.36	1.09
Northwest	0.57*	7.29*	1.35*	2.71*	0.68*	6.90*	1.91*	4.50*
West Midlands	0.02	3.70*	0.34	0.67	0.04	3.34*	0.49	1.47*
East Midlands	-0.01	3.54*	0.49*	0.47	-0.00	3.02*	0.70*	1.02
East Anglia	0.02	2.13	0.29	0.25	0.00	1.71	0.36	0.44
Greater London	0.11	0.38	0.63	-0.01	0.05	-0.27	0.88	-0.79
Southeast	0.04	0.58	0.31	0.05	0.01	0.19	0.38	-0.33

Metropolitan Borough	0.44*	1.62	0.58*	1.33*	0.48*	1.32	0.76*	1.30*
Inner London	0.47	3.19	0.43	1.12	0.41	2.33	0.45	0.97
Outer London	0.17	1.34	0.11	0.58	0.19	1.27	0.11	0.69
Metropolitan County	-0.12	-0.10	-0.04	-0.12	-0.08	-0.06	0.00	0.09
Nonmetropolitan Borough	0.07	0.05	0.08	0.23	0.08	-0.14	0.11	0.03
Conservative Incumbent	-0.05	1.05	-0.05	-0.07	-0.05	1.04	-0.07	0.02
Labour Incumbent	0.36*	2.71*	0.98*	1.65*	0.45*	2.57*	1.44*	2.37*
Liberal Incumbent	0.05	-3.89*	-1.31*	-0.75	0.02	-3.43*	-1.91*	-1.66
SDP Incumbent	-0.19	-5.23*	-1.11*	-0.13	-0.25	-4.98*	-1.66*	-2.69*
Liberal Candidate	0.15*	0.11	0.15	0.40*	0.17*	0.15	0.20	0.46
R^2 (adj.)	0.88	0.86	0.82	0.88	0.89	0.88	0.83	0.88

Table 8.6 Regression Analyses: Flows to Alliance, by Class

1979 Vote	Middle Class				Working Class			
	Conservative	Labour	Liberal	Did Not Vote	Conservative	Labour	Liberal	Did Not Vote
Constant	11.10	24.49	69.19	19.50	11.93	22.44	72.28	19.22
Liberal 1979	-0.03	0.25*	0.11*	0.06*	0.02	0.26*	0.17*	0.14*
Agriculture	-0.03	0.03	-0.06	-0.03	-0.03	0.02	-0.06	-0.01
Energy	0.04	-0.29*	-0.13*	-0.06	0.01	-0.29*	-0.24*	-0.15*
Manufacturing	0.04*	-0.03	0.03	0.02	0.03	-0.04	0.01	0.00
Unemployment	-0.02	-0.55*	-0.62*	-0.26*	-0.08*	-0.53*	-0.83*	-0.38*
Working Class O-O	-0.04*	0.00	-0.03	-0.02	-0.03	0.01	-0.01	-0.01
Marginality	0.02	0.09*	-0.03	-0.00	0.01	0.08*	-0.02	0.01
North	-0.67	-3.81*	-1.52	-1.37*	-0.74	-3.69*	-1.70	-2.13*
Yorks/Humberside	-0.94*	-4.60*	-3.45*	-2.25*	-1.20*	-4.43*	-3.78*	-2.96*
Northwest	-1.42*	-5.45*	-4.67*	-2.74*	-1.52*	-5.05*	-4.04*	-3.31*
West Midlands	-1.48*	-2.73*	-2.90*	-1.94*	-1.41*	-2.47*	-2.50*	-1.95*
East Midlands	-2.01*	-3.86*	-4.96*	-2.91*	-1.99*	-3.53*	-4.70*	-2.90*
East Anglia	-1.03	-2.39*	-2.67	-1.57*	-1.07	-2.17*	-2.66	-1.59*
Greater London	-2.05	-4.89	-7.90*	-4.01*	-2.53	-4.77	-8.69*	-4.26*
Southeast	-1.38*	-2.15*	-3.58*	-2.08	-1.47*	-2.02*	-3.58*	-2.01*

Metropolitan Borough	-0.48	-1.12	-0.87	-0.41	0.02	-1.06	-1.56	-0.86
Inner London	1.16	-2.76	-1.90	-0.59	0.71	-2.68	-3.14	-1.69
Outer London	0.66	-0.13	1.35	-0.58	0.68	-0.26	0.86	0.06
Metropolitan County	-0.06	-0.20	-0.20	-0.05	0.01	-0.26	-0.52	-0.21
Nonmetropolitan Borough	0.29	-0.65	-0.52	-0.16	0.19	-0.70	-1.00	-0.57
Conservative Incumbent	-0.48	-0.27	-0.33	-0.38	-0.35	-0.17	0.08	-0.18
Labour Incumbent	-0.53	-2.31*	-2.54*	-1.25*	-0.63	-2.19*	-3.12*	-1.58*
Liberal Incumbent	6.80*	4.65*	9.23*	6.48*	6.14*	3.93*	7.96*	4.45*
SDP Incumbent	3.69*	5.73*	7.04*	4.70*	3.56*	5.20*	6.74*	4.25*
Liberal Candidate	0.17	0.30	0.30	0.66	0.20	0.33	0.35	0.21
R^2 (adj.)	0.28	0.72	0.43	0.41	0.28	0.75	0.58	0.63

Table 8.7 Regression Analyses: Flows to Non-voting, by Class

1979 Vote	Middle Class				Working Class			
	Conservative	Labour	Liberal	Did Not Vote	Conservative	Labour	Liberal	Did Not Vote
Constant	12.58	21.36	11.84	23.58	18.30	27.77	15.09	40.01
Party 1979	-0.15*	-0.10*	-0.10*		-0.15*	-0.16*	-0.10*	
Agriculture	-0.09	-0.02	0.00	-0.10	-0.11	-0.01	0.02	-0.01
Energy	0.12*	-0.12*	0.12*	0.15*	0.16*	-0.14*	0.09	-0.19*
Manufacturing	0.01	-0.08*	-0.03	-0.02	-0.01	-0.09*	-0.05	-0.14*
Unemployment	0.56*	0.16*	0.69*	0.83*	0.76*	0.21*	0.68*	0.33*
Working Class O-O	-0.01	0.02	0.00	-0.01	-0.01	0.03	0.02	0.06*
Marginality	0.11*	0.17*	0.09*	0.14*	0.13*	0.20*	0.10*	0.21*
North	-0.26	-0.39	0.28	0.42	-0.13	-0.59	0.56	-0.48
Yorks/Humberside	1.51*	1.27*	2.54*	3.01*	2.06*	1.42*	2.90*	1.97*
Northwest	1.07*	-0.88	2.21*	1.10	1.51*	-1.06	2.51*	-0.44
West Midlands	0.53	0.56	1.20	0.79	0.76	0.64	1.66*	1.40
East Midlands	1.12*	1.38*	2.83*	2.21*	1.63*	1.58*	3.42*	2.60*
East Anglia	1.49*	0.83	1.77*	2.02*	1.75*	0.95	1.95*	1.08
Greater London	4.26*	4.17*	6.43*	6.95*	5.23*	4.68*	7.00*	4.75*
Southeast	1.59*	1.88*	2.50*	2.60*	2.04*	2.17*	2.83*	2.45*

Metropolitan Borough	1.58*	0.34	1.44*	2.30*	2.42*	0.43	1.45*	1.24*
Inner London	4.90*	1.72	3.86	6.27*	6.63*	2.04	4.01	4.25
Outer London	0.23	-0.38	-0.25	0.86	0.88	-0.34	-0.26	0.74
Metropolitan County	0.23	0.61	0.75	1.21	0.56	0.56	0.78	0.99
Nonmetropolitan Borough	0.89*	0.98*	1.05*	1.89*	1.44*	1.16*	1.16*	1.51*
Conservative Incumbent	-0.09	-0.05	-0.21	-0.42	-0.12	-0.07	-0.06	0.35
Labour Incumbent	0.25	-0.57	1.36*	0.82	0.50	-0.64	1.39*	-0.74
Liberal Incumbent	-0.98	-1.61	-3.81*	-1.25	-0.63	-1.81	-4.36*	-0.82
SDP Incumbent	0.22	-0.26	-3.36*	-0.69	0.10	-0.18	-3.89*	-1.01
Liberal Candidate	-0.20	-0.39	-0.39	-0.65*	-0.23	-0.44	-0.44	-0.45
R^2 (adj.)	0.82	0.71	0.63	0.71	0.80	0.76	0.58	0.62

constituencies with large percentages employed in the energy industries.

Finally, Table 8.7 reports the regression equations for the flows into the *non-voting* category. Again, the R^2 values are substantial, averaging 0.70. Of the groups of independent variables, only one — that relating to incumbency — comprises largely insignificant regression coefficients. The flow from voting Liberal in 1979 to non-voting in 1983 was significantly influenced by incumbency variables, however; not surprisingly, constituencies contested by Alliance incumbents had relatively small flows in this category, in both classes, whereas constituencies with Labour incumbents reported significantly large flows.

The two main hypotheses presented here (p. 212) to account for the geography of non-voting relate to party support in 1979 and marginality: both are entirely verified. The flow from each party into non-voting was less, the greater that party's support in the constituency in 1979; the general tendency was for voters to desert parties *where* they were weakest. Desertion was also greatest in the safer seats, as were the rates of continued non-voting; people were more likely to abstain, it seems, when their vote was likely to be irrelevant to the result. In addition, there was above-average movement into the non-voting category, and above-average continued non-voting, in constituencies with high unemployment levels. Of the other socio-economic characteristic variables, only that relating to the percentage employed in energy industries was generally significant. In the mining constituencies, apparently, above-average numbers of 1979 Conservative and Liberal voters, and of non-voters, failed to vote in 1983 among the middle class, whereas below-average percentages of Labour voters failed to register their vote. Among the working class in those constituencies, there were relatively small percentages of Labour voters switching to non-voting and of non-voters once again abstaining, but above-average percentages of working-class Conservatives failing to vote. Clearly, the Labour Party was able to retain support much better in such constituencies than it was in the country as a whole.

The switch to non-voting was also regionally patterned, with all but eight of the regression coefficients showing shifts into that category which were greater than those recorded in the Southwest. There was no clear north:south division, however; neither the North nor the West Midlands region was clearly different from the Southwest (except with the shift of working-class Liberals in 1979 in the latter) and there were significant differences for 1979 Conservative and Liberal voters only in the Northwest and in East Anglia. The most substantial regional difference was between Greater London and the Southwest, followed by the Southeast and the East Midlands. Perhaps the number of alienated voters was already high in the northern regions, and the shifts between 1979 and 1983 reflect growing disillusion with the political system in parts of the generally more prosperous southern regions. The inner cities and, even more, the nonmetropolitan boroughs (places like Bristol, Leicester and Nottingham) record significantly large non-voting flows.

Similarities and Differences

A general picture to be derived from the four tables just discussed is that although there were substantial absolute differences between the two classes in the flow-of-the-vote, the relative differences were few; where a working-class flow was small so, in most cases, was the middle-class flow, and vice versa.

The validity of this interpretation was checked, as in the previous two chapters, by using principal components analyses. These were run separately for each class, and also for the two classes combined, for each of the four categories of flow (those to Conservative, to Labour, and to the Alliance, and those involving non-voting at both elections). The results (Table 8.8) show that the generalisation above very clearly held for the Conservative and Labour Parties, across both classes, but less so for flows to the Alliance and to non-voting. There was, it seems, but one general geography of the flows to Conservative and Labour, common to both classes, which differed in their absolute

Table 8.8 Percentage of Variation Accounted For by the First Principal Component

Destination of Flows	Middle Class	Working Class	Both Classes
Conservative	90.9	90.5	90.5
Labour	96.5	96.7	96.4
Alliance	80.9	85.0	82.2
Did Not Vote	70.8	60.8	63.5

Table 8.9 Regression Analyses: Principal Components

Vote 1983	Conservative	Labour	Alliance	Did Not Vote
Constant	-0.72	-1.52	1.11	-1.97
Party 1979	0.04*	0.03*	0.03*	
Agriculture	0.01	-0.01	-0.00	-0.01
Energy	-0.03*	0.03*	-0.03*	0.03*
Manufacturing	-0.01	0.00	0.00	-0.00
Unemployment	-0.08*	0.05*	-0.09*	0.16*
Working Class O-O	0.00	0.01	-0.00	0.00
Marginality	-0.01*	-0.02*	0.00	0.01*
North	0.03	0.33*	-0.46*	-0.10
Yorks/Humberside	0.03	0.15	-0.68*	0.31*
Northwest	-0.05	0.47*	-0.83*	0.10
West Midlands	0.10	0.21*	-0.49*	0.02
East Midlands	0.11	0.20*	-0.76*	0.26*
East Anglia	-0.04	0.05	-0.41*	0.30*
Greater London	-0.40	0.16	-1.11*	0.84*
Southeast	-0.08	0.12	-0.52*	0.34*

Metropolitan Borough	-0.45*	0.39*	-0.16	0.41*
Inner London	-0.44	0.17	-0.30	0.71*
Outer London	-0.08	0.20	0.10	0.14
Metropolitan County	-0.15	0.15	-0.04	0.27*
Nonmetropolitan Borough	-0.23*	0.17*	-0.09	0.23*
Conservative Incumbent	0.15*	-0.05	-0.06	-0.13*
Labour Incumbent	-0.21*	0.41*	-0.40*	0.23*
Liberal Incumbent	0.12	-0.20	1.39*	-0.50*
SDP Incumbent	-0.50	-0.26*	1.16*	-0.22
Liberal Candidate	-0.02	0.04	0.05	-0.09
R^2 (adj.)	0.86	0.87	0.56	0.73

levels but not in their relative size across the 515 constituencies. There was a general pattern for each of the other two sets, too, but also substantial deviations from it.

These general patterns can be represented by the scores on the relevant principal components. Here, the scores for the analyses of the two classes combined (the final column of Table 8.8) have been used as the dependent variables in regression analyses. The results, which indicate the predictability of the geography of the flow-of-the-vote, are given in Table 8.9.

The regression for the flows to the Conservative Party contains only eight significant independent variables. More voters shifted to Conservative in 1983, or remained loyal to that party, in constituencies where the party was strong in 1979, where there were few energy workers and unemployed, and if the seat was marginal. In the more urban constituencies (the metropolitan and nonmetropolitan borough seats) the flow to Conservative was less, as it was in constituencies where a Labour incumbent was standing. A Conservative incumbent attracted an above-average flow from the other parties and from non-voting in 1979 and retained an above-average loyalty from 1979 Conservative voters. (Interestingly, although this is clearly a significant relationship, it does not appear in any of the eight patterns that the principal component synthesises; see Table 8.4.)

Comparing the first column of Table 8.8 with the regressions in Table 8.4 indicates where the particular flows deviate from the general. Noteworthy in such comparisons are the greater relative success of the Conservative candidates among the middle class in the more agricultural constituencies, the regional variations in the flow from Labour and non-voting in 1979 to Conservative in 1983, as well as in Inner London, and the impact of Alliance incumbents on certain flows.

Given that the principal component ccounts for just over 90 per cent of the variation in the eight maps, then the residuals from the regression of that component on the independent variables pick out the constituencies where the party did better and worse than predicted. Figure 8.5 maps the residuals, identifying those constituencies where the difference between the actual and predicted values was greater than one standard error of the predicted values. As would be expected from the high R^2 value for the regression (0.86), these residuals are widely scattered, with only small local concentrations. Among the

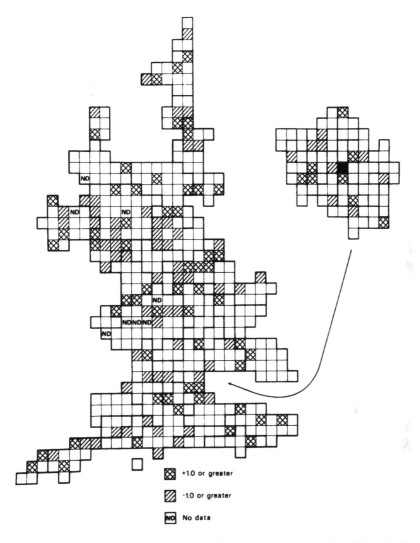

+1.0 or greater

-1.0 or greater

No data

Figure 8.5 Residuals from the Regression on the Principal Component for the Flows to Conservative, by Class

positive residuals — those where the flows to Conservative were substantially larger than estimated by the regression equation — are concentrations in: northern Tyne and Wear (Tyne Bridge, Tynemouth and Wallsend); south Durham (Darlington, Hartlepool and Stockton North); Dewsbury and Wakefield (the Conservative candidate won the former); north Nottinghamshire and west Lincolnshire (Brigg, Bassetlaw, Newark, Sherwood and

Nottingham North: two of these — Bassetlaw and Sherwood — are mining constituencies, indicating the relatively right-wing nature of the Dukeries coalfield (Waller, 1984a), whereas Nottingham North is the most working class of that city's constituencies: nearby Ashfield recorded a negative residual, however, reflecting the slow return of the constituency to the Labour dominance that preceded the by-election of 1977 — see p. 73); the growth centres of the M4 corridor — Basingstoke, Newbury and Swindon; north Cornwall; and west London along the Thames (Fulham, Putney and Richmond).

Among the clusters of negative residuals in Figure 8.5, where the Conservative flows were substantially overpredicted, are: much of Barnsley and Sheffield, where Labour is traditionally strong and has a stranglehold on local governments which were in conflict with central government over most of the period prior to the election; the west Lancashire coalfield (Leigh, Makerfield, St Helen's South and Wigan), presumably more radical than its Nottinghamshire counterpart; the Warrington area (reflecting the SDP near-success there in a by-election in 1981); Nottingham East and South (both of which the Conservative candidates nevertheless won); Oxford; and Brent. As with the positive residuals, these small clusters appear to indicate local circumstances producing flows that differed from the general pattern for constituencies in similar type locations and with similar circumstances.

The regression for the general component referring to flows to *Labour* (Table 8.9) also reproduces the salient elements of the eight separate regressions (Table 8.5). The Labour Party won more converts from those who did not vote Labour in 1979, and attracted more loyalty among 1979 Labour voters, in the constituencies where it was already strong, in the mining constituencies, in marginal contests, in areas with high unemployment, in the northern and Midland regions (except Yorkshire and Humberside), in the borough constituencies, and where it was fielding an incumbent: it did less well when facing an SDP incumbent (i.e. in all but one case, a former Labour MP).

The map of the major residuals (Figure 8.6) shows several local concentrations, some of them indicating considerable intra-regional heterogeneity. In the North region, for example, there are clusters of both negative (e.g. north Tyne and Wear) and positive (e.g. west Cumbria) residuals. The main clusters of

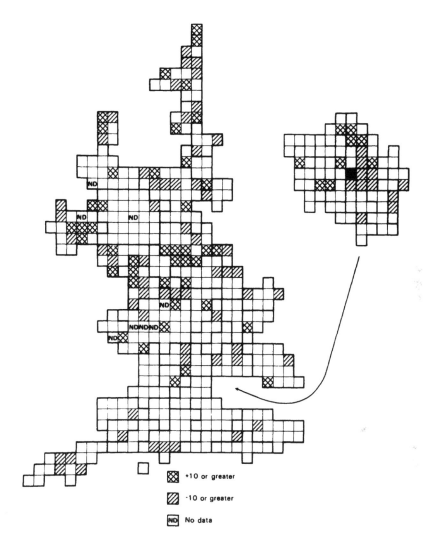

Figure 8.6 Residuals from the Regression on the Principal
Component for the Flows to Labour, by Class

positive residuals, where Labour attracted more votes than
predicted, are: the Sheffield area; parts of the west Lancashire
coalfield, extending into the Liverpool suburbs; the Potteries;
Battersea and Putney; and north London (including Hornsey,
Finchley and Tottenham). The local concentrations of negative
residuals — poor Labour vote-winning performances — are:
Plymouth; Gosport, Havant and Portsmouth (a local 'Falklands'

Factor' in a major naval base?); Hackney, Newham, Tower Hamlets and Southwark (including several constituencies with considerable strife within the party: Southwark was lost in a by-election a few months earlier); Nuneaton and Rugby; the Hertfordshire new towns (Hemel Hempstead, Stevenage and Welwyn); the Notts.-Lincs. border; and south Staffordshire.

For flows to the *Alliance*, the regression in Table 8.9 repeats the salient features of those in Table 8.6. The fits are low, however, (an R^2 of 0.56 for the general pattern) suggesting that there was a substantial amount of place-particular movement to the Alliance, that should be clearly identifiable in the map of residuals. Figure 8.7 identifies these.

In the south, outside London, the above-predicted concentrations of Alliance vote-winning include: Plymouth (one of whose seats was held by David Owen); Yeovil (another Alliance success) plus the nearby constituencies of Tiverton and Wells; parts of north Kent (Faversham, Gillingham and Maidstone); much of Hertfordshire (where the Labour vote declined substantially); and Chelmsford (where the Liberal candidate just failed to unseat the Conservative incumbent, Norman St John Stevas) and the adjacent seat of Colchester South. Within London, the Alliance did better than average in: the far southwest (Kingston, Richmond and Twickenham); the far east (Bexleyheath, Erith and Woolwich; the last of these was won by the SDP incumbent); and a ring of seats around the City (Bethnal Green, Bow, Islington South, Southwark and Vauxhall), in an area where the Liberal Party remained strong throughout the 1920s. It did less well than predicted in Newham, Hackney, the City and Westminster North, and in Ealing, Feltham and Hammersmith; outside London, the southern concentrations of poor Alliance vote-winning included much of Bristol, Slough and East Berkshire.

Further north, several concentrations of Alliance above-average performance in winning converts include constituencies held by incumbents. They include Crosby (Shirley Williams) and nearby Bootle, and Stockton South (Ian Wrigglesworth) — but not Stockton North (Bill Rodgers) — Langbaurgh and Redcar, plus Warrington (a near-win by Roy Jenkins in 1981). (Note, however, the absence of such clusters around Newcastle upon Tyne Central (J. Horam) and East (Mike Thomas).) Further south, the Alliance did well in south Humberside (Glanford and

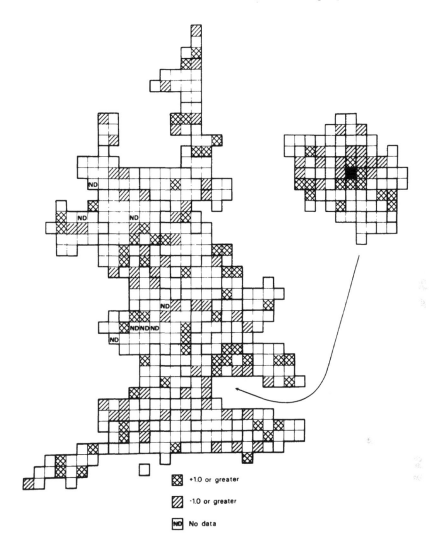

Figure 8.7 Residuals from the Regression on the Principal Component for the Flows to Alliance, by Class

Scunthorpe plus Great Grimsby; note that this is an area where Steed and Curtice (1983) dispute the BBC/ITN estimates), central Lincolnshire (East Lindsey and Gainsborough), and in east Coventry and Nuneaton; the flows to the Alliance were much less than predicted in Leicester.

Finally, the regression for the flows into the *non-voting* category general component (the final column of Table 8.9)

provides another very good fit, especially given the absence of any 1979 party vote variable (which varies among the flows). It was in the safe seats, in the mining constituencies, in the areas of high unemployment, in the cities and in the southeastern regions of England that above-average proportions either shifted from voting in 1979 to not voting in 1983 or did not vote in either contest.

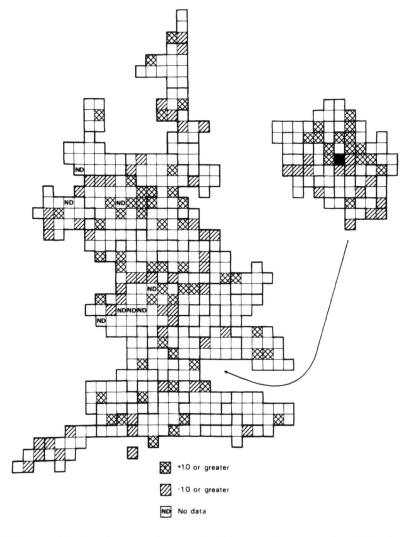

Figure 8.8 Residuals from the Regression on the Principal Component for the Flows to Non-voting, by Class

The residuals from this regression (Figure 8.8) identify some clear local nodes of either above-average or below-average shifts into non-voting. In London, for example, there is a substantial block of constituencies north of the Thames where non-voting was much more 'popular' than estimated by the general trends, whereas south of the river there were two blocks where it was decidedly 'unpopular'. Shifting into the non-voting category was unpopular in Plymouth, too, perhaps contributing to David Owen's success; the Liberal win in Yeovil was achieved against a larger-than-expected shift into the category that includes abstentions, however. The Midlands also contained clusters of positive (central Birmingham; Stoke-on-Trent; Leicester) and negative (Coventry; south Staffordshire) residuals. Pennine Greater Manchester contained several constituencies where non-voting was a popular option (Ashton, Littleborough, Oldham, Rochdale and Stalybridge: the Yorkshire constituency of Colne Valley seems to have been affected by this too), whereas in parts of Bolton and Bury the shift into non-voting was less than expected.

The Geography of Deviant Flows

The previous sections have identified the general geography of the flow-of-the-vote matrix, by class, and have pointed to the possible reasons for the main deviations from that geography. Some individual constituencies appear to have peculiar characteristics on one set of flows only; others stand out as not conforming on any set. To identify the latter more precisely, a map has been prepared showing these 'major deviants' (Figure 8.9); as in an earlier exercise (Figure 6.21) this has been compiled as the number of appearances for each constituency in the large residual category (greater than plus or minus one standard error) in each of the four preceding figures (8.5-8.8).

Thirteen constituencies stand out in Figure 8.9 as having four substantial residuals, one in each of the analyses reported above; their flows all deviated substantially from what was predicted by their socio-economic characteristics, location and electoral context. Two of the northern constituencies — Tynemouth and Boothferry — fell into this category by having larger-than-expected flows to Conservative (relative to regional, etc. norms) and smaller-than-expected flows to all other three (Labour, Alliance, non-voting). Sedgefield, Liverpool Broad Green and Crewe, on the other hand, had larger-than-expected flows to

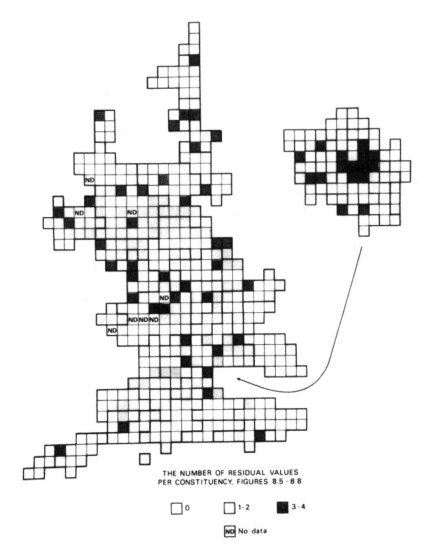

THE NUMBER OF RESIDUAL VALUES
PER CONSTITUENCY, FIGURES 8.5 - 8.8

□ 0   □ 1 - 2   ■ 3 - 4

ND No data

Figure 8.9 The Geography of 'Deviant' Flows: the Number
of Residual Values per Constituency in Figures 8.5-8.8

Labour and non-voting countered by smaller-than-expected to
Conservative and Alliance. Crewe was held by the Labour
incumbent, Gwynneth Dunwoody, by only 290 votes, whereas
the BBC/ITN estimates suggested a Conservative victory —
clearly an added personal factor in one of the few Labour seats in
the area; Liverpool Broad Green was one of the few constit-

uencies where the Alliance Parties found it impossible to agree on a joint candidate, with an independent Liberal standing against the official Alliance candidate — Richard Crawshaw, an SDP incumbent — and getting 7,000 votes to the latter's 5,000 in a seat which the BBC/ITN estimates suggested that Conservative should win, but Labour did. Leyton, too — a seat lost by an SDP incumbent, Bryan Magee — had larger-than-expected flows to Labour and non-voting; in Wigan, the Conservative vote collapsed and there were larger-than-expected flows to both Labour and the Alliance.

The other constituencies in this group of 'deviants' where the Alliance benefited from larger-than-expected flows were: Great Grimsby, where all others experienced smaller-than-expected flows; Plymouth, Devonport, where the Alliance was the only beneficiary of larger-than-expected flows, presumably a reflection of the local popularity of David Owen; Bermondsey, where Simon Owen built on his success in the by-election a few months earlier, while the Conservatives also received a substantial boost and the Labour Party, for which this had been a safe seat with over 63 per cent of the votes in the BBC/ITN estimates, suffering smaller-than-expected flows; and Yeovil, where the unexpected Alliance success was apparently founded in part on larger-than-expected flows into the non-voting category. Finally, in two London constituencies — Hackney South and Putney — the Conservative Party experienced larger-than-expected flows, along with non-voting in the former (where the Conservative candidate nevertheless came a poor second) and Labour in the latter (very much a marginal seat, fought by Peter Hain for Labour).

None of these 13 constituencies is in the list of eight identified as the major deviants from the class-based analyses of the 1983 voting in Chapter 6. This suggests that it was particular, very local influences which caused the deviations in the flows. Indeed, many of these deviations seem to indicate the influence of one or more individual candidates, and there is little evidence in Figure 8.9 that any of the 13 is at the centre of a block of 'deviating constituencies'; Hackney South and Leyton are adjacent but deviated in different directions. The same conclusion can be drawn regarding several of the constituencies with three residuals recorded in Figure 8.9, although not always in the same direction: Shirley Williams achieved a larger-than-expected flow to the Alliance, relative to the Northwest norm and the 1979 result, in Crosby, for example, whereas the flow to the Alliance was smaller-than-expected in the Rochdale seat retained for the Alliance by Cyril Smith. Patricia Hewitt apparently succeeded in attracting a larger-than-expected flow to Labour in Leicester East (although she did not win the seat), as did Claire Short in Birmingham, Ladywood and Dennis Howell (the former Minister for Sport and 'Environmental Disasters') in Small Heath. (Their neighbour, Jeff Rooker held on to Perry Bar despite a larger-than-expected flow to Conservative but aided by a similar flow to non-voting and a smaller-than-expected flow to the Alliance.)

Finally with regard to Figure 8.9, one part of the country stands out as having very few deviant constituencies. In East Anglia, virtually every constituency went along with the national and regional trends. In Norfolk, only in the North-West constituency, whose incumbent was the sole Conservative MP to defect to the SDP — Christopher Brocklebank-Fowler — was there more than one deviant flow. (Interestingly, these were smaller-than-expected for Conservative and Labour and larger-than-expected for non-voting.) In Suffolk, only in Ipswich, somewhat unexpectedly held by Ken Weetch for Labour with a larger-than-expected flow to his party and a smaller-than-expected flow to the Alliance, was there any substantial deviation from the general pattern. Cambridgeshire had no constituency with more than one residual. Essex, classified here as one of the Home Counties of the Southeast region, shares these characteristics with the East Anglian counties: only in Billericay and Colchester South were there two deviations. Perhaps these East Anglian constituencies lacked individual candidates able to 'buck' the national and regional trends?

In Summary

This chapter has rounded out the analyses of Chapters 6 and 7. It has shown substantial differences between the middle and working classes in the flow-of-the-vote at the national scale, and has shown very great spatial variability in the various elements of the flow, especially for the Labour Party. For both Conservative and Labour, the geographies of the various flows terminating in those parties are very similar in relative terms, suggesting that where members of one class go — in terms of voting — so do members of the other. The similarities were less for the Alliance and non-voting, though still substantial, indicating geographical variability in the appeal of these voting options, depending on starting point (1979 vote) and place.

As throughout the analyses in this book, the geographical variability in the flows was very predictable in terms of the statistical models applied; only with the flows from Conservative to Alliance was there real failure. This gives further support to the hypotheses advanced here, which are the focus of the summary discussion in Chapter 10.

9 CAMPAIGN SPENDING AND VOTES

One element of a general election has been excluded from this analysis — the campaign itself, on the doorsteps and in the homes. Most electioneering in England takes place in the mass media — radio, TV and the newspapers — and the competing parties seek to influence the electorate by their national presentation of both policies and personalities. Most of this campaigning for votes has no explicit spatial element; the parties are seeking to influence the electorate as a whole, and hoping that votes will be translated into seats. (Johnston, 1979b, p. 117 gives a, non-English, example of spatial campaigning.) At the local level, the candidates and the party-workers build on the national campaign, setting it in the local context and directing it at the target groups of voters.

The dominance of the national campaign is perhaps a further reason why British electoral analysts focus so much on national trends and pay relatively little attention to local campaigning; in most analyses, local activity is at best used to account for the residuals from the national trend. But there is a great deal of local activity. Some of it has the goal of convincing voters that a given party-cum-candidate is best for them. But most of it is concerned with the more mundane task of canvassing — of ensuring that the electors are aware of the candidates and issues, of identifying who a party's likely supporters are, and of encouraging a high level of turnout among those supporters.

The few studies that have been made of local canvassing in England suggest that a well-executed strategy can be extremely effective (see the review in Taylor and Johnston, 1979, pp. 326ff.). Nevertheless, general studies (e.g. Rose 1967, 1976) either ignore the local effort entirely or suggest that it has very little influence.

The detailed analyses of the impact of local canvassing and campaign efforts have focused on individual constituencies (or parts thereof), and have been concerned with the impact of canvassing on both turnout and the result. Clearly, such an analysis is not feasible in a study of voting trends in 523 constituencies, and it is necessary to employ a surrogate variable. One such surrogate is the level of expenditure in each constituency by each party during the pre-election campaign period, and the possible effect of such spending is examined in the rest of this brief chapter.

Spending and Votes

Much of the work done at the constituency level in the campaign period costs the parties nothing. It is undertaken by volunteers, who address envelopes, visit voters on their doorsteps, and so on. It is impossible to measure this activity. It is possible, however, to obtain a precise measure of the amount of money spent, and to estimate its impact.

The Representation of the People Act (1983) limits the amount of money that can be spent by each candidate — other than the candidate's personal expenses — during the campaign period. In 1983 these limits (Section 76) were:

£2,700 in a county constituency, plus 3.1 pence for every entry on the electoral roll; and
£2,700 in a borough constituency, plus 2.3 pence for every entry on the electoral roll.

Thus the amount that can be spent is a function of the size of the electorate — the more electors, the greater the costs of informing them; more can be spent per elector in county than in borough constituencies because of the greater costs of reaching the electorate in lower density areas.

In total, £4.9m was spent in England during the 1983 campaign, as follows:

	£
Fees to agents	245,466
Fees to clerks and messengers	91,031
Printing, advertising, stationery, etc.	3,856,704
Expenses of public meetings	82,904
Hire of committee rooms	147,003
Miscellaneous	458,470

(All of the data here and analysed in this chapter are from the official election returns: *Election Expenses*, 1983.) Most of the spending, therefore, (79 per cent) was on printing and advertising, on producing posters and leaflets with which to inform the electorate about the party's candidate and policies. (The parties are allowed to send one item of electoral literature to each elector through the post, free, as long as it is personally addressed.)

Almost £5m is not a small sum for the three political parties combined. At the 1979 general election, for example, it is estimated that the Conservative Party spent £2.3m from its central office, whereas Labour spent £1.6m and the Liberal Party £154,000 (Pinto-Duschinsky, 1981a, pp. 145, 167, 205). Relatively little of this is transferred to the constituencies (the maximum quoted by Pinto-Duschinsky is 16 per cent, by Labour), and the local parties rely almost entirely on their own efforts (subscriptions, donations and fund-raising) to obtain the money to finance their campaign (largely the printing of election leaflets). Their expenditure then was £2.8m, suggesting that the parties raised about £2m locally, to be spent on informing and convincing the local electorate.

According to these figures for 1979 (the latest election for which all data are available), between 40 and 50 per cent of all expenditure by the political parties during the general election campaign took place at the constituency level. Very little attention has been paid to this expenditure, however. The Committee on Financial Aid to Political Parties (Houghton, 1976) made a detailed survey of the state of party finances but undertook no analyses of the impact of party spending. Pinto-Duschinsky (1981a, 1981b), similarly, reports on the level of constituency spending but does not assess its impact.

In contrast to the situation in the UK, a great deal of work has been done on the efficacy of campaign spending in the United States (the most comprehensive study is Jacobson, 1980). There spending is much larger (Jacobson reports an average of $90,000 for incumbent Republican Representatives in 1976, for example, and $880,000 for incumbent Republican Senators) and the constraints on spending are much looser — TV time can be bought, for example. The result is that there are significant relationships between spending and success; money obtains votes (e.g. Caldeira and Patterson, 1982), although the relationships are not always straightforward, as Jacobson's sophisticated analyses show.

These American analyses have stimulated some British work (e.g. Johnston, 1979a, 1983e). The results have been largely negative, although there was some evidence that the Liberal Party obtained significant returns from its expenditure at both the October 1974 general election and the 1979 European Assembly election. (See also Johnston, 1984c.) These analyses have also been criticised (Gordon and Whiteley, 1980), in part because of assumed deficiencies in the data. As Pinto-Duschinsky (1981a, p. 249) notes, 'The law is not watertight' and election agents can manipulate it to some extent. Nevertheless, especially with the costs of printing, etc. comprising the majority of the expenditure, there are limits to this, and the reported expenses can be taken as reasonable surrogate indicators of the amount of canvassing, etc. effort in each constituency. It seemed worthwhile, therefore, to round off the analyses in the previous three chapters with a brief assessment of the impact of spending at the constituency level on the 1983 election result.

Hypotheses

The simplest hypothesis that could be advanced is that the greater the amount spent by a party locally (as a percentage of the allowed expenditure there) the greater the percentage of the votes it should get. With three parties contesting every seat, this implies that, *ceteris paribus*, the more a party spends the more votes it should get and the more its opponents spend the less votes it should get.

This simple hypothesis needs to be modified, however. As suggested elsewhere (Johnston, 1983a) and indicated here, continuity characterises the geography of voting in England. Thus the pattern of spending should reflect this continuity, with the parties spending most in the constituencies where they feel spending is likely to have most influence (this recursive model is developed by Jacobson, 1980). These are clearly the marginal constituencies, so analysts should be looking for (a) whether the pattern of spending is rational in its direction, and (b) whether any deviations from such rationality influence the pattern of votes (for the 1960s, see Johnston, 1984c).

A two-stage model is thus generated:

(1) The more marginal the constituency, the greater the party expenditure; and
(2) the more that a party's expenditure deviates from this norm, the greater its impact (positive or negative) on the result.

Thus in an extremely marginal seat, one would expect all three parties to spend the maximum allowed, and that this expenditure would have no effect. If one party spent less than the others, however, it would suffer in terms of votes won.

This simple model assumes not only perfect rationality on behalf of the parties — each can assess the situation in each constituency exactly (remembering that the 1983 election was not being fought in the 1979 constituencies and that the Alliance was introducing a new element to the political scene) — but also that the constituency parties were able to raise the needed money. Neither assumption is particularly valid: the parties vary spatially in their strengths, money-raising abilities and assessment of the situation. To account for the geography of spending, therefore, it may be necessary to incorporate other variables. For example, the parties may be better able to raise money to defend seats that they hold. They may also decide that certain constituencies are 'lost causes' — because of a high unemployment rate in the case of the Conservatives, for example. Thus the analyses here are exploratory. They focus (a) on variations in the level of spending, and (b) on the impact of variations in spending on the cells of the flow-of-the-vote matrix analysed in Chapter 8.

Who Spent What, Where?

The data for the 515 constituencies analysed here (excluding the eight with no Liberal candidate in 1979: p. 164) show that on average Conservative candidates spent 76.3 per cent of the allowed maximum (standard deviation, 20.7), Labour candidates spent 65. per cent (24.5), and Alliance candidates spent 57.5 per cent (26.4). The implication is that the frequency distributions are negatively skewed. Figure 9.1 shows that this is certainly the case with Conservative spending; in more than half of the constituencies this was 81 per cent or more of the maximum, but in about one-tenth the party spent less than 40 per cent of the allowed total. For both Labour and the Alliance the distributions are more rectangular, with Labour's slightly bimodal.

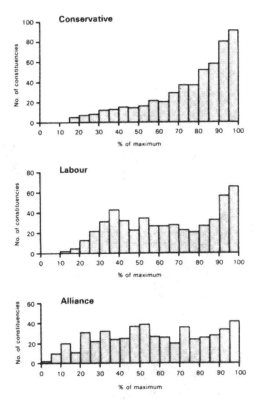

Figure 9.1 Frequency Distributions of Campaign
Expenditure as a Percentage of the Allowed Maximum

Figures 9.2-9.4 present the geography of these frequency distributions, by quartiles. In general terms, both the Conservative and the Labour Parties seem to have spent most in their main areas of strength; Conservative in the south and in rural areas, Labour in the north and in London. There are clear deviations from this, however. The Conservative Party spent above-average sums in several of the Alliance-held seats in the north, for example (e.g. Berwick, Crosby and Rochdale) but it was much less forthcoming with expenditure in the Alliance's nodes of support in east London. (It did spend above-average sums in Croydon North West and in Richmond, however.) Labour was especially active in London, perhaps because of greater local affluence, and raised and spent no more than average in some of its strongholds, such as South Yorkshire and the Potteries. The Alliance, too, was very active in London,

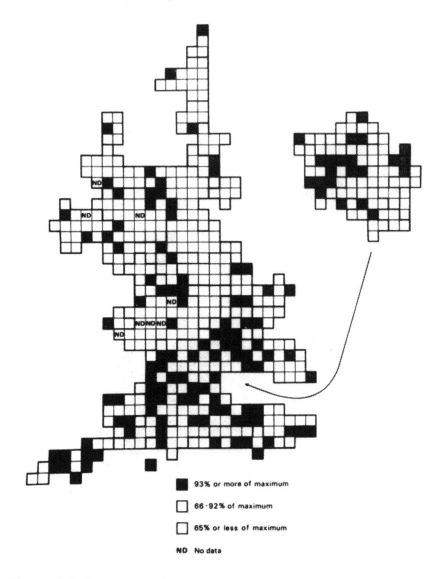

Figure 9.2 Campaign Expenditure by the Conservative Party

countered by low levels of spending throughout much of the Midlands.

The overall impression of these three maps is that they are very dissimilar. If the rationality argument held, then one would expect very high correlations among the maps, since each party should be spending most where the result is in greatest doubt.

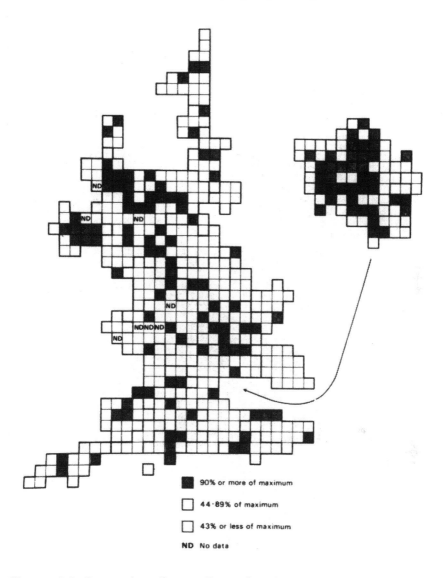

Figure 9.3 Campaign Expenditure by the Labour Party

Those correlations did not occur, however. Instead, the computed relationships were

Conservative : Labour	−0.04
Conservative : Alliance	0.41
Labour : Alliance	0.01

81% or more of maximum

37-80% of maximum

36% or less of maximum

ND No data

Figure 9.4 Campaign Expenditure by the Alliance

Each party, it seems, pursued a separate spending policy (or, as is more likely, each of the three sets of local parties had a different geography of money-raising ability and, perhaps, of decisions where money was most needed).

What, then, were the correlates of the spending? It has already been suggested that parties would spend most (a) where they held the seat and (b) where the outcome of the election was most in doubt. The first of these influences was readily included in statistical models by dummy variables indicating which party 'held' the seat according to the BBC/ITN (1983) estimates for 1979. The second relates to marginality. A variable measuring this — the distance between the two leading parties in the estimated 1979 result — has been used throughout the analyses in this book; the expectation here is that there should be a negative relationship between that measure and spending. This marginality measure refers only to the two leading parties. One would expect that where all three parties had a strong chance of winning, then spending would be even greater. To measure 'three party marginality' the simple entropy index (see Johnston, 1976, for an application) was used whereby

$$H = \sum_{i=1}^{3} p_i \log_2 \frac{1}{p_i}$$

where

p_i is the proportion of the votes won by party i in the 1979 election.

(Since proportions were being used and i was 3 for every constituency, no standardisation or transformation to an information measure — see Johnston and Semple, 1984 — was needed.) As H increases, so the proportions become more equal, and H reaches its maximum when $p_1 = p_2 = p_3$; at this stage, according to the general hypotheses about the geography of spending, campaign expenditure by the parties should be at a maximum.

Three sets of regressions were conducted, therefore, using the same 515 constituencies analysed in Chapters 7 and 8. The dependent variables were the percentages of the maximum spent in each constituency by the Conservative, Labour and Alliance Parties. At the first stage in each set, the spending variables were regressed against the two marginality variables. (The correlation between these two was only −0.34.) At the second stage, two dummy variables representing 1979 'victories' were introduced; at the third, the full battery of socio-economic, regional, constituency type and incumbency variables was added.

Table 9.1 Regression Analyses of Spending: First Stages

	Conservative	Labour	Alliance
Stage 1			
Constant	−76.59	199.96	−87.80
Marginality	−0.18*	−1.32*	−0.07
Entropy	49.87*	−33.72*	46.73*
R² (adj.)	0.18	0.39	0.09
Stage 2			
Constant	39.83	150.19	−37.61
Marginality	−0.54*	−1.21*	−0.08
Entropy	10.14*	−19.69*	28.95*
Conservative Victory	25.85*	–	16.27*
Labour Victory	–	9.20*	–
Liberal Victory	30.21*	−21.94*	−7.06
R² (adj.)	0.45	0.42	0.28

The results of the first two stages are shown in Table 9.1. They provide at best only weak support for the general hypotheses regarding rationality in spending. The *Conservative* Party was most rational, in terms of both goodness-of-fit and the signs of the regression coefficients. More was spent in the more marginal seats, especially those with the strongest three-party challenges, and more was spent in seats notionally held by that party relative to those held by Labour (in the constant term). In addition, substantially more was spent in those seats notionally held by the Liberal Party (five only) than in the Labour seats. Nevertheless, more than half of the variation in Conservative spending remained unaccounted-for.

For the *Labour* Party, the intriguing element of the results is the apparent avoidance of spending in the constituencies with the closest three-party contests; although Labour spent more in the marginal seats defined by the distance between the two main parties, it spent less in those where all three parties were relatively equal according to the 1979 estimates. As expected, it

spent more in the constituencies that it held than in those held by
Conservative — though not substantially so — but it spent
significantly less in Liberal-held than in Conservative-held seats.
Finally, *Alliance* spending was much less predictable. Perhaps not
surprisingly, there was no significant relationship with the two-
party marginality measure, but a strong and significant one with
the three-party. Compared to Labour-held seats, the Alliance
spent more in Conservative-held, suggesting that it perceived
these as more likely to yield electoral returns; on average, it
spent less (though insignificantly so) in the five constituencies that
it notionally won in 1979 than in the 193 notionally held by
Labour.

Turning to the final stage, the goodness-of-fit increases
substantially in each case, though in none is more than 53 per
cent of the variation in spending accounted for (Table 9.2). For
the *Conservative* Party, most money was spent defending
Conservative-held seats, with special attention to those in which
the margin of success over the next party was small. In addition,
less was spent in all regions outside the Southwest and Southeast,
including Greater London, and the seats with large percentages
of energy and manufacturing workers attracted below-average
spending. In general, therefore, the party was better able to raise
campaign funds in its traditional heartlands, and it spent
rationally in terms of protecting its seats, especially those most
under threat. The *Labour* Party, too, concentrated spending on
the seats most under threat but not on those that it held. It spent
more countering SDP than Liberal Alliance candidates and in the
constituencies with both large percentages of manufacturing
workers and of unemployed, with less in the more agricultural
constituencies and those with the higher levels of embourgeoisement. Again, this appears to be reflecting both the party's
strengths — where it is best able to raise money — and a rational
allocation of funds. The main exception is its low average level of
spending in the West Midlands, a region which many commentators thought held the key to the election because of recent
industrial decline.

The *Alliance*, too, was very much concerned with the more
winnable seats — on both measures. Like the Conservative
Party, it also raised and spent more where it had seats to defend
— not those which it notionally won in 1979 but those in which it
was fielding incumbents, most of whom were defectors to the

Table 9.2 Regression Analyses of Spending: Final Stage

	Conservative	Labour	Alliance
Constant	82.46	79.81	29.08
Marginality	−0.62*	−0.99*	−0.23*
Entropy	5.90	−1.44	25.28*
Conservative Victory	17.57*	–	1.14
Labour Victory	–	−4.83	–
Liberal Victory	16.95	−30.05*	−19.72
Agriculture	0.11	−0.87*	1.14*
Energy	−0.84*	0.33	−1.28*
Manufacturing	−0.31*	0.36*	−0.65*
Unemployed	−0.54	1.22*	−0.66
Working Class O-O	−0.05	−0.40*	−0.44*
North	−10.43*	3.76	−16.11*
Yorks/Humberside	−6.80*	0.43	−6.49
Northwest	−9.90*	1.41	−9.78*
West Midlands	−3.66	−7.87*	−10.23*
East Midlands	−4.04	−2.25	−12.96*
East Anglia	−2.84	5.82	−8.71
Greater London	−21.67*	10.94	−20.34
Southeast	0.16	−0.24	−7.24*
Metropolitan Borough	−2.89	5.28	−1.76
Inner London	10.47	9.89	16.09
Outer London	11.90	2.83	20.50
Metropolitan County	−3.80	−8.62	−7.00
Nonmetropolitan Borough	−1.01	4.78	4.46
Conservative Incumbent	−1.65	−0.06	−0.11
Labour Incumbent	0.55	2.23	−0.68
Liberal Incumbent	5.16	13.93	37.79*
SDP Incumbent	0.88	5.56	29.18*
Liberal Candidate	0.88	−3.53*	−13.25*
R^2 (adj.)	0.51	0.53	0.46

SDP. In other seats, the SDP appears to have been better able than its senior partner to raise and spend. There were significant regional variations, too, with low average spending in most of the northern and Midland regions. More was spent in the agricultural areas, but less in the solid working-class constituencies, with large workforces in mining, power and manufacturing, and also in those with relatively large percentages of working-class owner-occupiers.

Overall, these results suggest that the parties were in general sensible in their spending on the campaign, focusing on the seats where potential victory had to be defended and strong challenges thwarted. But there were considerable variations between types (in terms of socio-economic characteristics) and regional locations of constituencies in how much was spent, presumably reflecting variations in how much could be raised locally as a consequence of local party vitality.

And the Electoral Impact?

Did this varied volume of spending affect the distribution of votes? If more spending means better campaigning then, *ceteris paribus*, the more that a party spent the greater its ability both to retain its 1979 voters' loyalty and to attract support from those who voted otherwise then. Countering this, the more its opponents spent the less support it should retain and win.

To test whether this was so, the three variables representing party spending were added to the regressions of the flow-of-the-vote by class, reported in Tables 8.4-8.6. Reported here are the adjusted R^2 values from those analyses, the R^2 values with the three spending variables added, and the regression coefficients for the spending variables (Table 9.3). In brief, it appears that spending did have a significant, though not substantial, impact on the distribution of votes among the three parties, very largely in the expected direction.

For the *Conservative* Party, the addition of the spending variables increased the R^2 value by only 0.1 in six of the eight regressions, and by 0.3 in the other two (those involving flows from Liberal in 1979 to Conservative in 1983). In each of the eight equations, Conservative spending had a positive and significant impact; the more that the party spent, the better its performance. Labour spending significantly influenced Conservative performance only with regard to flows between the two; the more that Labour spent, *ceteris paribus*, the less successful the Conservative Party was in winning former Labour voters' support. Similarly, the more that the Alliance spent, the fewer 1979 Liberal voters who preferred the Conservatives in 1983. In

Table 9.3 The Impact of Spending on the Flow-of-the-Vote, by Class

Middle Class

1979	Con	Lab	Lib	DNV
1983	Con	Con	Con	Con
R² (adj.)	0.81	0.88	0.64	0.85
R² (adj.) with spending	0.82	0.89	0.67	0.86
Conservative	0.05*	0.02*	0.03*	0.05*
Labour	−0.01	−0.01*	0.00	−0.01
Alliance	−0.02*	0.01*	−0.03*	−0.01

1979	Con	Lab	Lib	DNV
1983	Lab	Lab	Lab	Lab
R² (adj.)	0.88	0.86	0.82	0.88
R² (adj.) with spending	0.89	0.88	0.83	0.88
Conservative	0.00	0.03*	0.01*	0.01*
Labour	0.01*	0.06*	0.01*	0.02*
Alliance	−0.00	−0.06*	−0.01*	−0.01*

Working Class

1979	Con	Lab	Lib	DNV
1983	Con	Con	Con	Con
R² (adj.)	0.81	0.87	0.63	0.85
R² (adj.) with spending	0.82	0.88	0.66	0.86
Conservative	0.06*	0.01*	0.02*	0.02*
Labour	−0.01	−0.01*	0.00	−0.01
Alliance	−0.01	0.00	−0.01*	0.00

1979	Con	Lab	Lib	DNV
1983	Lab	Lab	Lab	Lab
R² (adj.)	0.89	0.88	0.83	0.88
R² (adj.) with spending	0.90	0.89	0.84	0.89
Conservative	0.00	0.03*	0.02*	0.03*
Labour	0.01*	0.06*	0.01*	0.04*
Alliance	−0.00	−0.06*	−0.02*	−0.03

	1979				1983			
	Con All	Lab All	Lib All	DNV All	Con All	Lab All	Lib All	DNV All
R² (adj.)	0.28	0.72	0.43	0.41	0.28	0.75	0.58	0.63
R² (adj.) with spending	0.39	0.76	0.50	0.48	0.38	0.78	0.63	0.67
Conservative	−0.03*	−0.01	−0.05*	−0.03*	−0.03*	−0.01	−0.04*	−0.01
Labour	0.00	−0.04*	−0.02	−0.01	−0.00	−0.04*	−0.02	−0.01
Alliance	0.03*	0.07*	0.09*	0.05*	0.04*	0.06*	0.08*	0.07*

addition, among the middle class the more the Alliance spent the smaller the number of loyal Conservativs but the greater the number of Labour voters who transferred their allegiance to the Conservative Party.

For the *Labour* Party, too, the spending variables added on average less than 0.02 to the value of R^2. Only five of the 24 regression coefficients were insignificant, but overall the impact of spending was slight. In every case, the more that Labour spent, the better its performance, and the more that the Alliance spent, the poorer the outcome for Labour (though not significantly so for flows from Conservative to Labour). None of this was unexpected. What were entirely unanticipated were the six significant, positive regression coefficients for Conservative spending. The more that the Conservatives spent, the greater the loyalty to Labour and the greater the flows of 1979 Liberals and non-voters to the Labour camp. The most plausible explanation for this is that Conservative spending stimulated high levels of turnout by the more pro-Labour voters, as a negative reaction to the campaign material; knowing more about the opposition encouraged votes against.

For the *Alliance*, spending had a much greater impact, increasing one R^2 value by more than 0.10. As expected, the more that the Alliance spent, the better its performance, and the more that its opponents spent, the less satisfactory the outcome for the Alliance. Labour spending influenced only the flow of 1979 Labour voters to the Alliance; Conservative spending influenced the flow of non-voters and Conservative voters, plus Liberal loyalty. The Alliance did best where it spent a lot on getting its message across locally and its opponents, especially the Conservative Party, spent little.

In the previous section, it was shown that the level of spending in each constituency by the various parties could be predicted by many of the same variables included in the analyses reported in Table 9.2. It could be, therefore, that the significant relationships with the spending variables are spurious, reflecting collinearity with the variables that successfully predict the flow-of-the-vote. To test whether this was so, the spending variables were replaced by the residuals from the regressions reported in Table 9.2, so that the flow-of-the-vote was being regressed against that volume of spending not accounted-for by the various constituency characteristics. The results (Table 9.4) are very similar to those in

Table 9.3, indicating that the findings in the latter are almost certainly 'real': campaign spending did have a marginal, significant impact on the flow-of-the-vote to the various parties.

And the Non-Voters in 1983?

What of non-voting in 1983? Was it the case that the less that a party spent the more likely that its 1979 supporters did not vote in 1983? To test this, the spending variables (in raw form and as residuals from the regressions in Table 9.2) were entered into the regressions reported in Table 8.7. The results are given in Table 9.5.

Again, the regression results show small, but significant and 'real', relationships between spending and votes. All of the significant coefficients (12 of the 24 in each analysis) are in the expected direction, showing that the more that was spent on campaigning, the smaller the flows into, and 'loyalty' within, the non-voting category. Overall, Alliance spending had the greatest impact, with only the flows from Labour to non-voting not being influenced. Labour spending influenced only the shift from a Labour vote in 1979 to non-voting in 1983 plus the level of non-voting 'loyalty' among the working class; similarly, Conservative spending influenced only the flows from that party, plus non-voting 'loyalty' among the middle class. (The findings of class differences in which party influenced non-voting loyalty suggest that Conservative and Labour campaigning had most impact within their traditional class support.)

In Summary

This brief chapter has not attempted an exhaustive study of campaigning in 1983, but has focused on the more limited topic of the impact of campaign spending in the constituencies — largely on printing campaign literature and posters — on the voting. It has shown that such spending had a significant, though small, impact on the distribution of votes. Where a party spent more, so more of its 1979 supporters remained loyal, *ceteris paribus*, and fewer either transferred their allegiance to the other parties or failed to vote.

Table 9.4 The Impact of Spending on the Flow-of-the-Vote, by Class: Spending Variable Residuals

Middle Class

1979	Con	Lab	Lib	DNV	Con	Lab	Lib	DNV
1983	Con	Con	Con	Con	Lab	Lab	Lab	Lab
R (adj.)	0.81	0.88	0.64	0.85	0.88	0.86	0.82	0.88
R² (adj.) with spending	0.82	0.88	0.67	0.86	0.89	0.88	0.83	0.88
Conservative	0.05*	0.02*	0.03*	0.04*	0.00	0.03	0.01*	0.01*
Labour	-0.00	-0.00	0.01	0.00	0.01*	0.06*	0.01*	0.02*
Alliance	-0.02*	0.01	-0.03*	-0.01	-0.00	-0.06*	-0.01*	-0.01*

Working Class

1979	Con	Lab	Lib	DNV	Con	Lab	Lib	DNV
1983	Con	Con	Con	Con	Lab	Lab	Lab	Lab
R (adj.)	0.81	0.87	0.63	0.85	0.89	0.88	0.83	0.88
R² (adj.) with spending	0.81	0.87	0.66	0.85	0.89	0.89	0.84	0.89
Conservative	0.05*	0.01*	0.02*	0.02*	0.00	0.03*	0.02*	0.02*
Labour	-0.00	-0.01	0.00	-0.00	0.01*	0.06*	0.01*	0.04*
Alliance	-0.01	0.00	-0.01*	0.00	-0.00	-0.06*	-0.01*	-0.02*

	Con All	Lab All	Lib All	DNV All	Con All	Lab All	Lib All	DNV All
1979								
1983								
R^2 (adj.)	0.28	0.72	0.43	0.41	0.28	0.75	0.58	0.63
R^2 (adj.) with spending	0.38	0.76	0.49	0.48	0.37	0.78	0.62	0.67
Conservative	−0.03*	−0.03*	−0.05*	−0.03*	−0.03*	−0.01	−0.04*	−0.01
Labour	0.00	−0.02	−0.02	−0.01	−0.00	−0.03*	−0.01	−0.01
Alliance	0.04*	0.07*	0.08*	0.05*	0.04*	0.06*	0.08*	0.07*

Table 9.5 The Impact of Spending on the Flows into Non-voting, by Class

	Middle Class				Working Class			
1979 1983	Con DNV	Lab DNV	Lib DNV	DNV DNV	Con DNV	Lab DNV	Lib DNV	DNV DNV
R^2 (adj.)	0.82	0.71	0.63	0.71	0.80	0.76	0.58	0.62
Raw Spending Variables								
R^2 (adj.) with spending	0.82	0.72	0.65	0.73	0.81	0.77	0.61	0.63
Conservative	−0.02*	−0.01	0.01	−0.04*	−0.03*	−0.01	0.00	−0.01
Labour	0.01	−0.02*	0.01	−0.01	0.00	−0.03*	0.00	−0.03*
Alliance	−0.01*	−0.01	−0.04*	−0.02*	−0.01*	−0.01	−0.05*	−0.02*
Residual Spending Variables								
R^2 (adj.) with spending	0.82	0.72	0.65	0.72	0.80	0.77	0.61	0.63
Conservative	−0.02*	−0.01	0.01	−0.03*	−0.02*	−0.01	0.01	−0.01
Labour	0.00	−0.03*	0.00	−0.01	0.00	−0.03*	0.00	−0.03*
Alliance	−0.01*	−0.01	−0.04*	−0.02*	−0.02*	−0.01	−0.05*	−0.02*

Of the three parties, it seems that the Alliance got the greatest return from spending. In part this reflects the much greater variation among constituencies in the amount spent by the Alliance than by the Conservative Party; Labour spending was almost as variable as the Alliance's, though at a slightly higher level, but its impact was more restricted. The Alliance, of course, had a new message to sell, and it was perhaps because of this that the greater the marketing effort the greater the impact. Money may not 'buy' votes as it does in the United States, but spending it in the constituency campaign certainly does no harm, and may bring in a small — perhaps very important — return.

10 SUMMARY AND CONCLUSIONS

The goal of this book has been to demonstrate the importance of the geographical element in the study of voting at the 1983 general election in England. This goal has been more than adequately achieved. Demonstration is insufficient, however; what is vital is understanding. Thus this summary seeks both to identify the salient features of the research reported here and to chart a possible route towards their understanding.

The Results in Précis

It is commonplace in British electoral analysis to treat England as a homogeneous place, with no spatial variations in voting behaviour additional to those that can be accounted for by what are considered to be the basic electoral cleavages. Thus, for example, if national surveys showed that 55 per cent of the working class voted Labour, and 22 per cent of the middle class, then it would be assumed that these percentages held in each component area of England; differences in the Labour percentage of the vote between places would merely represent differences in the class composition of the electorate. Similarly, it is generally assumed that if 10 per cent of the Labour voters at one election switched to the Alliance at the next, then that percentage applied in each and every part of the country. Some analysts have accepted that this is not so, but their recognition of geographical variations in voting is not fully reflected in either popular or general academic treatments of English electoral behaviour.

The analyses in Chapter 2 have shown, beyond any reasonable doubt, that these assumptions simply were not valid with regard to the 1983 general election result in England. The national patterns of voting by occupation, class and housing tenure did not produce reasonable approximations of the distribution of the votes between Conservative, Labour and Alliance. Nor did the national flow-of-the-vote matrix, when applied to the (estimated) 1979 distribution of votes in the constituencies, produce reasonable estimates of the 1983 result. In brief, it must be concluded that

Voting in the English constituencies in 1983 was not a series of local representations of the national trends.

By itself, this conclusion provides the foundation for detailed investigations of reasons for the geography of voting then. The analyses in Chapter 3 provide further justification for such work. They show that the geographical variations from the national trends identified in Chapter 2 produced an election result, in terms of the allocation of seats among the parties, which was very different from what would have occurred if the national trends had been repeated in every constituency. Thus the nature of the British government after June 1983 was very much influenced by the geography of voting on 9 June.

Given these findings, the remainder of the book set out to uncover the main dimensions of the geography of the voting. Chapter 4 provided the guidance for these explorations, with a review of other studies of similar phenomena. This suggested a variety of local influences on voting, which are often compounded under the umbrella title of 'structural' or 'neighbourhood effects'. These, it was suggested, can be unravelled into the following types: contagious effects; sectional effects; environmental effects; and campaign effects. Various hypotheses relating to these were developed, focusing on the last three, and means of testing them were outlined in Chapter 5. The results of the analyses in Chapters 6-9 are summarised here.

Sectional Effects

A sectional effect is a long-standing geographical element of voting whereby differences in local and regional political culture produce spatial variations in the support given to the various political parties. These variations are not independent of the class and other cleavages on which English electoral behaviour is based — in all areas, for example, members of the working class are much more likely to vote Labour than are members of the middle class. But the spatial variations mean that within any class there are substantial differences in their propensity to vote for a particular party, reflecting the local environment.

Because the analyses reported here relate only to voting in 1983 and to changes in voting between 1979 and 1983, the longevity of such sectional effects could not be investigated. All that could be tested was whether: (a) there were variations in the support given to each party which reflected the strength of the party's base in the constituency; (b) there were, in addition, variations between different types of constituency; and (c) there were also variations reflecting particular socio-economic characteristics in the constituency.

Regarding the *geography of voting in 1983*, the analyses showed:

(i) For both the Conservative and the Labour Parties, the stronger the social base of support in the constituency, the greater the vote for the party, in all social classes. (For Conservative, the base of support was middle-class owner-occupied households; for Labour, it was working-class households.)

(ii) For the Conservative Party, the three northern regions (North; Yorkshire/Humberside; Northwest) provided below-average support, in all social classes, whatever the constituency base; for Labour, those three regions provided above-average support, as did Greater London.

(iii) The Conservative Party obtained below-average support, from all social classes, in the metropolitan borough and county constituencies and also in the nonmetropolitan boroughs. Relatively, its strength was in the rural areas, and in addition the Conservative vote increased the larger the percentage of the constituency workforce employed in agriculture; the Labour vote fell, in all social classes, as the percentage employed in agriculture increased.

(iv) The Labour Party obtained above-average levels of support, from all classes, in the metropolitan borough constituencies (i.e. the inner cities). Its vote also increased in ratio to the percentage employed in both energy (the mining and power industries) and manufacturing; the Conservative vote fell as the percentage employed in energy increased (suggesting strong pro-Labour sentiments in the mining constituencies), but its vote also increased (though at a much smaller ratio than was the case for

Labour) as the percentage in manufacturing increased.

(v) For the two Alliance parties, there was a major regional component to the voting, with support being greatest in the Southwest — among all social classes — and significantly less elsewhere. The lowest overall average performance was in Greater London.

(vi) There was no clear urban:rural division to the Alliance vote; however, the Alliance did better among the working class, the larger the percentage of the workforce employed in agriculture, and less well, among all classes, the larger the percentage employed in energy industries. Thus its regional base was apparently strengthened among the rural working class.

(vii) Non-voting was greatest among the middle classes where they formed a small minority of the electorate, whereas it was greatest among the working class where they were in a substantial majority.

(viii) Non-voting was lowest on average, within all social classes, in the Southwest region, and was significantly greater in almost all cases elsewhere. It was greatest in Greater London and in the metropolitan constituencies, but its level declined as the percentages in both agriculture and manufacturing increased.

Regarding the *geography of voting shifts, 1979-83*, the analyses showed:

(i) For all three parties, the larger their percentage of the vote in a constituency in 1979:
    (a) the larger the percentage of their voters then who remained loyal in 1983; and
    (b) the larger the flows to them of voters who expressed different preferences in 1979.

(ii) The Conservative Party retained more support, and won more from other sources, the greater the percentage of the workforce employed in agriculture and, in general, the smaller the percentage employed in energy industries.

(iii) The Labour Party retained more support, and won more converts, the greater the percentage of the constituency electorate employed in energy.

(iv) The Conservative Party was, on average, relatively unsuccessful at both retaining support and at winning converts in the urban constituencies.

(v) In general, the Labour Party retained more support in the three northern regions; it was also relatively successful at winning converts in the Midlands.

(vi) The Southwest region provided the Alliance with the highest levels of both voter loyalty (1979 Liberal voters) and converts. Its performance elsewhere was in most cases significantly lower.

(vii) The weaker a party's performance in a constituency in 1979, the larger the percentage of its voters then who did not vote in 1983.

(viii) The flows into, and within, the non-voting category were in general highest in the urban constituencies; they were lowest in the Southwest, North, and West Midlands.

Together, these results provide a great deal of support for the hypothesised sectional effects. Both Conservative and Labour performed better in constituencies where their socio-economic base was stronger, among all social classes, than in those where it was weaker; only the Alliance, which in any case was seeking to establish a non-class-base, did not obtain more votes where its major supporters were more numerous. Conservative, Labour, and the Alliance did better, again among all social classes, where they were strong in 1979. This is substantial circumstantial evidence of continuity in voting patterns, at the constituency level. It was accentuated in certain types of constituency; the Conservative vote was greater in the more agricultural, whereas Labour did better in those with more power and mining workers.

The individual constituencies occupy locations within the British space-economy, and the analyses have shown that relative location is one part of the sectional effects; voting in a constituency reflects not only the characteristics of the electorate but also the context within which they live and work. There were, for example, clear inter-regional and metropolitan:urban:rural contrasts in voting, which were additional to those related to the constituency characteristics. England, it seems, is divided into a series of sections (north:south; urban:rural; London and metro-politan:other urban; inner city:suburbia; etc.) within which voters are apparently socialised. It is possible, therefore, to reach two general conclusions from this part of the analysis:

The geography of voting Conservative and Labour in 1983, by all social classes, reflected inter-regional and urban:rural divisions within England, and was in addition closely related to the relative size of the party's social base in each constituency. For the Alliance, the major sectional influence operated at the regional scale.

and

The geography of voting shifts between 1979 and 1983 showed that Conservative, Labour and Alliance were all better able to retain support of 1979 voters, and to win the support of those who voted otherwise then, in the constituencies where they were strong in 1979. In addition, the Conservative Party performed relatively badly in urban constituencies, Labour did relatively well in the northern regions and the Alliance's best vote-winning performances were in the Southwest.

Thus the established sectional strengths were maintained and the inter-regional and urban:rural polarisation of English voting behaviour increased.

For the Conservative and Labour Parties, these sectional effects were by far the major component of the statistical 'explanation' for both the geography of voting by class and the geography of the flow-of-the-vote by class. For the regression analyses of the component scores for the former (Table 6.5) the single social class variable (middle-class owner-occupiers for Conservative; working class for Labour) accounted for 66 per cent of the variation in the Conservative vote and 64 per cent of the variation in the Labour vote; for the Alliance, the relevant variable (social class II) accounted for only 9 per cent. And with the flows, using the scores (Table 8.9) as dependent variables, 73 per cent in the variation of the flows to Conservative was associated with the strength of the Conservative vote in 1979, and the relevant percentage for flows to Labour, regressed on the 1979 Labour vote, was 75. For the Alliance, the percentage was 33. Thus for Conservative and Labour the sectional effect reflecting each party's traditional heartland was the strongest; for the Alliance, it was the inter-regional variability in its appeal that stood out.

Environmental Effects

Environmental effects reflect the salience of local issues, which produce deviations from the general pattern of voting. They may be brought about by either the salience of particular issues in the constituency or by the personal qualities of a party's candidate.

Regarding *issue effects*, two were focused on here — the local (i.e. constituency) level of unemployment and the sale of homes to working-class occupiers. The first of these increased substantially during the four-years' incumbency of the previous government, and to the opposition parties that government's policies were the major cause of increased unemployment. It was anticipated, therefore, that the Conservative Party would suffer in areas of high unemployment whereas more electors there would either vote Labour or Alliance or, because of increased alienation from governments, abstain. On the second issue, the Conservative Party sought to increase working-class owner-occupancy through the sale of council houses; where this policy was most effective it should have reaped more votes. The analyses showed:

(i) That the Conservative vote declined, among all social classes, the higher the rate of unemployment, whereas the Labour vote and the rate of non-voting increased. The Alliance vote was also lower the higher the level of unemployment, suggesting a further process of electoral polarisation in England: Conservative and Alliance performed best in the more prosperous areas; Labour performed better in the less prosperous constituencies, where in addition more voters opted out of voting at all.

(ii) The higher the level of unemployment in a constituency, the lower the loyalty to both Conservative and Alliance, in all social classes, the greater the loyalty to Labour, the greater the shifts to Labour, the smaller the shifts to Conservative and Alliance, and the greater the flows into the non-voting category. The more prosperous:less prosperous electoral cleavage was accentuated over the four-year period.

(iii) Regarding the issue of working-class owner-occupancy, the results are not readily interpreted. Within each social class and tenure group, the larger the percentage of working-class households in a constituency who were owner-occupiers, the smaller the Conservative and Alliance vote, and yet those parties should have benefited most from embourgeoisement.

(iv) The percentage of working-class owner-occupiers in each constituency had no apparent influence on the flow-of-the-vote between 1979 and 1983.

It can be concluded, therefore, that with regard to the variables included in the models tested here, the only local issue which significantly influenced the geography of voting, over the country as a whole, was the level of unemployment. (Of course, there may have been many other issues which were entirely local, in that they influenced one or a few constituencies only. The analysis here looked only at national issues which varied locally in their salience.) Thus

The greater the level of unemployment in a constituency the lower the popularity of Conservative and Alliance in 1983, and the poorer the performance of these parties relative to 1979. Against this, Labour popularity was greater, and increased relative to 1979, and in addition more electors failed to vote (seemingly delivering their verdict by abstaining).

Regarding *candidate effects* the main analysis was of the impact of incumbency. In addition, the relative impact of Liberal and SDP candidates within the Alliance was explored. The results showed that, relative to non-incumbent candidates in each party:

(i) Conservative incumbents on average won more votes from all social classes and negatively influenced the performance of Labour candidates. They had no apparent influence on the flow-of-the-vote, however.

(ii) Labour incumbents on average won more votes from all social classes and negatively influenced the performance of Conservative and Alliance candidates. Labour incumbents also retained above-average support from 1979 Labour voters, attracted above-average flows from other voting categories in 1979, and in general reduced the flows to both Conservative and the Alliance.

(iii) Liberal incumbents had very substantial influences on the 1983 vote, increasing it by 14-15 per cent on average relative to the performance of non-incumbents and reducing very significantly the Conservative, Labour and non-voting percentages. They also retained the loyalty of substantially above-average percentages of the 1973 Liberal voters and attracted large flows to the Alliance.

(iv) SDP incumbents also recorded above-average votes for the

Alliance in 1983, though much less than their Liberal counter-parts (about 4 percentage points on average across all social classes); they also produced significantly lower voting for the Labour Party. Similarly, SDP incumbents on average retained the support of more 1979 Liberal voters and won over more converts, but again less than did their Liberal counterparts.

In addition,

(v) Liberal candidates on average won one percentage point more votes for the Alliance than did SDP candidates, in all social classes.

It is possible to conclude, therefore, that

> *For all parties, incumbency brought a bonus of votes: within the Alliance, Liberal candidates brought a larger bonus than did SDP candidates.*

All of these environmental effects were analysed with regard to the country as a whole. Inspection of the maps of residuals from the various regressions (summarised in Figures 6.21 and 8.9) identified other local effects. Some of these reflected the influence of particular candidates, and also of recent by-elections (as in Warrington where the SDP leader lost in 1981 and did not fight either seat in 1983, but whose organisation and goodwill clearly remained for his successors). Others identified areas which deviated significantly from expected trends, given their locations and constituency characteristics: they included the relatively conservative coal-mining area of north Nottingham-shire, for example, and the strength of Labour in both South Yorkshire/north Derbyshire and the Potteries. Many of these, it seems, are nodes of local political culture which either accentuate or counter the general trends.

Campaign Effects

This final set of effects refers to the local electoral context. It was incorporated in the analyses of Chapters 6-8 through the variable representing (estimated) marginality in 1979 and in Chapter 9 by a brief analysis of the impact of local campaign spending. The results showed:

(i) That the safer the seat, the smaller the Conservative and Labour votes, the larger the support for the Alliance, and the greater the non-voting percentage.

(ii) That the safer the seat, the lower the loyalty of 1979 voters to Conservative and Labour, the smaller the flows of voters to those parties, and the larger the shift to non-voting. In addition, the safer the seat the larger the shift from Labour to Alliance.

(ii) The more that each party spent in the local campaign the greater the loyalty of its 1979 voters, and the more voters that it won support from.

Thus

> *Vote-switching was more common in the safer seats and the longer-established parties (Conservative and Labour) did better in the marginal constituencies. The level of campaign spending had a small but significant impact on the result in each constituency.*

In general terms, this conclusion fits into the others with regard to long-term continuity in English voting behaviour. The socio-spatial patterns were most at risk where the result of the election was least problematic.

And the Reasons

English electors are influenced, according to theorists and analysts, to vote for the party which is most likely to promote policies in their supporters' self-interests (see, for example, Robertson, 1976). Thus the Conservative Party, with its emphasis on the free market, on personal property rights and on a meritocratic society, is associated with the middle class and with owner-occupiers, whereas the Labour Party and its greater adherence to state controls of the market, to collective consumption, equality of opportunity and equality of wealth and income, is linked to the working class and to tenants. The Alliance, which seeks to promote policies in the interests of all and without a class bias, has no clear socio-economic base.

Political and electoral surveys have shown the accuracy of these links and illustrate the important role of the socio-economic cleavages in English electoral politics. Those conducted at the time of the 1983 general election produced no exceptions, except that support for the Alliance was greater than that for the Liberal Party in previous decades. The analyses reported here, however, have indicated that the analyses of national data present an over-simplified view of English electoral behaviour. Using a relatively novel estimating procedure, it has been shown that support for a party within a socio-economic group varied very substantially across the 523 English constituencies, as did changes in that support between 1979 and 1983. There is undoubtedly an English political culture, which is clearly represented in the national surveys, but there are also major local variations around the main lineaments of that culture. A full analysis of electoral behaviour in England must account for those local variations.

The statistical analyses presented here have been extremely successful, in their own terms (i.e. the values of R^2), at providing accounts for those local variations. They suggest that each party is stronger than might be expected in the areas where its electoral base is surest, and that this strength is maintained against the vicissitudes of national trends. Such local strengths are set in a matrix of regional strengths; in addition the Conservative Party performed better in the more rural areas and the Labour Party in the mining communities. Thus the social bases to voting in England interact with sectional bases, to produce a socio-spatial set of electoral cleavages. These display long-term continuity, though issues that are locally salient (e.g. race in the 1960s, unemployment in 1983), particular candidates, and the nature of the local campaign can all produce deviations from that continuity.

The sectional, environmental and campaign effects are all elements of the geography of voting. What is unclear is how they came about, especially the sectional effects. Why were members of the working class more likely to vote Conservative in a middle-class dominated constituency in the rural Southeast than were their counterparts in northern, working-class dominated, mining constituencies? Is it because they are not directly comparable, despite their overt similarity, or is it because they are influenced by their environments?

Most of the attempts to account for these differences focus on the environmental influence, and many present this influence as operating through the medium of inter-personal exchange —

what are generally known as contagious effects. The greater the weight of opinion favouring a certain point of view in an area, it is claimed, the greater the likelihood that all voters, whatever their social position, will be exposed to it personally and will accept that point of view. In voting, people are influenced by their neighbours, so that where there is a local majority for a certain point of view, that majority position is accentuated.

The results presented here are consistent with that model of voting behaviour; in the 1983 general election, both the Conservative and the Labour Parties won more votes amongst all social groups in the areas where their support was greatest and all three parties were best able to retain support in 1983 in the constituencies where they were strong in 1979. Voters, it seemed, talked with their neighbours, and voted accordingly. Or did they?

Two arguments can be made against this model. The first is that it is hard to verify; as indicated in Chapter 4, there is little evidence of contagious effects in operation. Further, some authors (such as Dunleavy; see also Harris, 1984), doubt the whole basis of the contagious effect model. Whatever the balance of classes in a constituency, most people live among others of the same class, and have most of their social contacts with them. Voters live in locally-segregated class 'ghettos', whatever the wider situation. Secondly, the model focuses almost entirely on the individual voters, choosing among the political parties. Very little attention is paid to the activities of those parties, as they seek to build up support, not only in the pre-election period but also in a continuous process of political education, involved in local as well as national government and in a wide range of other arenas (see Taylor, 1984a, 1984b). And yet it is the parties that structure the agenda of politics. As Schattschneider (1960) expresses it

> Democracy is a competitive political system in which compet-
> ing leaders and organizations define the alternatives of public
> policy in such a way that the public can participate in the
> decision-making process (p. 141)

In doing this, the organisations (i.e. the parties) both control the agenda — 'the definition of the alternatives is the supreme instrument of power' (p. 68) — and seek to ally voters to particular positions with respect to the agenda — 'organization is the mobilization of bias' (p. 71).

Much of this mobilisation in English politics is class-based and national. But it is also place-based, if for no other reason than that the British electoral system demands it. Parties, if they are to win power, must win seats and therefore need spatial concentrations of support; as Gudgin and Taylor (1979) have shown, and has been demonstrated here for 1983, certain frequency distributions of votes across constituencies are better than others in order to fulfil the goal of winning Parliamentary representation and power.

Of the three parties which contested the 1983 general election in England, Labour has done most to develop a socio-spatial electoral base; it has stood the party in good stead. The Conservative Party has always presented itself as a 'national' party, although it has to some extent developed a socio-spatial base as a consequence of Labour's success in the northern regions and in parts of London. The Alliance, too, has presented itself through a national appeal, with neither class nor sectional cleavage. In this, it largely succeeded — though not well enough to win seats — although it clearly also benefited from the traditional regional (non-conformist) base of Liberal support, especially in the Southwest.

The electoral map which comes through repeatedly in the analyses here has persisted for more than 60 years (Johnston, 1983a). Indeed, there has been remarkable continuity in the electoral geography of England despite, as Taylor (1983) makes clear, four clearly-defined political 'epochs' in the period 1933 to 1983: the politics of national interest (to 1945); the politics of social democratic consensus (1945-59); the politics of technocracy and disillusion (1959-74); and the politics of crisis (1974 on). The Conservative and Labour Parties have dominated throughout these four epochs; their policies have been changed as reactions to the contemporary situation, but their electoral base has not and the parties have retained the same socio-spatial support.

Why has this happened? Why have the two parties been able to mobilise the same segments of the population throughout this period, producing a sequence of similar electoral geographies that was not dented by the 'breaking the mould' attempts of the Alliance in 1983? Given that the Labour Party has been the most successful in forging a clear regional electoral base, it could be that the electoral geography reflects the success of Labour, when in power, in representing the interests of the electorate in certain

regions (basically the depressed industrial areas of the north —
plus central Scotland and South Wales, which are excluded from
the analysis here) and winning benefits for them. Superficially
this would seem to be the case — the Labour Party has been
more generous with regional aid programmes in recent decades,
for example, and has offered aid to more areas than the
Conservative Party (see the maps in Dicken, 1983), which has
focused its attention recently on the major unemployment
blackspots only (Townsend, 1980) — with no apparent electoral
return!

There is, however, little evidence that the Labour Party has
successfully advanced the cause of the areas where, increasingly
according to the 1983 results, most of its electoral support is
based. Sharpe (1982) has argued that

> as the party of the underdog the Labour Party ought also to be
> the party of the periphery and of the localities (p. 135)

countering the immense centralisation in British politics and
society and seeking to remove the geography of inequality. But it
has not: despite the growing spatial polarisation of the British
electorate, the Labour Party has neither placed the removal of
the geography of inequality as central to its manifesto nor
countered the centralist trends in British society. (It did, in the
late 1970s, promote — in the end, unsuccessfully — devolution to
Scotland and Wales, largely to protect its electoral base there
against national incursions: see Bulpitt, 1983; Rose, 1982b.) Nor
has the Labour Party advanced the autonomy of local governments. Like the Conservative Party, it too has promoted a
centralist set of policies though not as draconian as those
introduced by the Conservatives in the early 1980s. To Sharpe,
this is a major puzzle.

Linked to this puzzle is the question 'why, if Labour has not
promoted the interests of its electoral base, has it not lost the
support of voters in the peripheral regions?' (This could be
expressed alternatively as 'why has the pro-Labour environmental effect in certain parts of England remained for so long?')
The answer, it is suggested, lies in the way the Labour Party in its
formative decades built a very firm electoral base in the industrial
regions (more so in some — Sheffield and the Potteries, for
example — than others — such as north Nottinghamshire).

Labour became the 'natural' party of local government there, as well as of Parliamentary representation, and its close links with the trade union movement associated it with the politics of the workplace and the home (with the unions acting as friendly societies before the development of the welfare state after 1945) as well as with the politics of the ballot box. The workforce was organised around Labour, and new voters — women who gained the vote in the 1920s and young people thereafter — were socialised into an environment in which Labour dominated the political scene.

According to most studies, it is the home that provides the major focus of political socialisation (see Butler and Stokes, 1974; Himmelweit *et al.*, 1981). The home is set in a wider social milieu, however, and where the two conform the influence of the home environment is likely to be greater than in situations where political tendencies in the home are countered by a majority view outside. Thus in the industrial regions where Labour built its electoral base in the 1920s, the strength of pro-Labour partisan attitudes created an environment which still remains. Further, success in these places could often only be achieved through a combination of social mobility (via education) and spatial mobility. The out-migration of the relatively successful, in particular to London and the Southeast, removed some of the potential local erosive forces and protected Labour's electoral base.

Voters decide which party to support at a particular election according to the salient issues then and the various party positions on these. It is possible that such 'issue-voting', with its potential for deserting traditional party allegiances, is more likely to occur in places where no one party has an established hegemony (as in 1945). Thus Himmelweit *et al.* (1981) claim that

> The degree of class-linked voting is time-bound and election specific except in socially homogeneous areas (the industrial North or the rural South) with a long and unbroken tradition of voting Labour or Conservative (p. 197)

The greater success of the Alliance in 1983, relative to Labour, in the south of England may reflect such a process; Labour's nodes of support there have declined substantially in recent decades, though it is unclear whether this is because they are isolated from the main Labour heartlands or because such places (e.g. Swindon) have benefited substantially from recent industrial expansion with weaker links to the traditional, Labour-based trade unions.

Himmelweit *et al.* (1981) also follow other electoral analysts (e.g. Sarlvik and Crewe, 1983) in arguing that issue-voting will increase in the future and that individual voters will display much weaker ties to particular parties — 'In the future the influence of the individual's past habit of voting on his or her subsequent vote will, if anything, decrease further' (p. 194). This process — widely known as dealignment (see Johnston, 1982a) — suggests that the geography of election results should become more variable, as people become more volatile in their partisan choices, unless voters' interpretations of the issues are strongly influenced by their local environmental cues (i.e. by contagious effects).

The 1983 general election result produced little evidence of the geographical variability following the greater individual volatility. Conservative and Labour both retained strength in their traditional heartlands. The Alliance, too, built on the Liberal base in certain areas — both at a regional scale (the Southwest), and around local nodes (in parts of London, for example, reflecting earlier successes, and in widely-scattered places such as Crosby, Rochdale and the Isle of Ely). The current geography of voting in England, in simple terms, comprises Labour strength in the north and parts of London, Conservative strength in much of the Home Counties and East Anglia, and the Alliance contesting Conservative control in parts of the south.

Given the nature of the British electoral system, a geographical base is highly desirable for the parties. Labour has a very well-developed socio-spatial foundation, without which 1983 would have been disastrous for it. That base is retreating, however, as Labour loses support around the edges and in the outliers.

The Conservative Party, too, has a firm socio-spatial base. The Alliance has not yet built one. Unless it does, it has little hope of electoral success without a massive shift of voter opinion in its favour. (The Alliance parties are fully committed to electoral reform, of course, which if implemented would remove the need for a socio-spatial base: Johnston, 1985. They are unlikely to achieve that goal unless they first win a substantial presence in the House of Commons under the present electoral system, however, so that they need a socio-spatial base in order to implement a system that dispenses with the need for a socio-spatial base!)

For all parties, a firm foundation in the local government system will almost certainly bring benefits in terms of votes and seats at general elections. This is widely recognised. In the Labour Party, for example, the important likely impact of local government reform on both local politics and the redefinition of Parliamentary constituencies was realised in the 1980s, and influenced several actions (see Johnston, 1979b, pp. 180-3). The Conservative Party, too, is well aware of the potential for 'gerrymandering' local government for electoral gains; its reorganisation of both Greater London (1963) and the rest of England and Wales (1972) placed nearly 58 per cent of the population in local authority areas (the counties outside the metropolitan counties, the districts within the metropolitan counties and Greater London) that were likely to be under continuous Conservative control, and only 17 per cent in those likely to be under continuous Labour control (Dunleavy, 1980c, p. 138). And, finally, the Liberal Party has been able to build on local electoral activity, even to the extent of winning Parliamentary seats — as in Liverpool, Bermondsey and Yeovil, where it was successful in 1983, and Chelmsford and Richmond, where it nearly was.

Conclusions

England is treated as a homogeneous spatial entity in many studies, not only of its electoral geography, with little or no recognition of internal cultural variations (except, perhaps, with reference to the 'Celtic fringe' of the Southwest). And yet the country has a rich cultural mosaic, as illustrated by studies of dialects and folklore. The differences between the various cultural regions are not sharp, but they are none the less 'real', and of relevance to a full understanding of many aspects of economy and society. This chapter has suggested that local political attitudes and beliefs, as structured by parties and other organisations, are part of that cultural mosaic.

England cannot be treated as a single unit with regard to

understanding voting behaviour there. The parties which have dominated English politics for the last half-century — Conservative and Labour — have developed not simply social cleavages within the electorate (whereby certain classes support one party and others support the second) but *socio-spatial cleavages*, with members of particular classes varying in their degree of support for a particular party according to its importance in the local political, social and economic milieu. Thus Labour gets its strongest support from the working class, on a national basis, but its support from all classes varies spatially — being greatest in those areas where working-class voters are most numerous and where Labour has traditionally been strong (in the area as a whole, and not just the constituency). Similarly, the Conservative Party gains most votes from the middle class nationally, but obtains its strongest support, from all classes, in constituencies and areas where middle-class milieux dominate. The Liberal Party has no firm class base (nor does its new ally, the SDP); it has made electoral inroads in certain areas — often on a pragmatic, *ad hoc* basis, at by-elections and via local government electoral success — and then has built on this support, establishing some spatial, if not socio-spatial, continuity of voting strength.

The analyses of the 1983 general election in England confirm this interpretation, presenting the electoral geography of England as the result of a continuing process of interaction between parties and voters (rather than simply a geography of voter choice, which is the usual approach). The technical sophistication has allowed the detail of the socio-spatial cleavages to be uncovered, and has clearly demonstrated the importance of the geographical perspective.

Within the overall continuity, change occurs, as voters react — in their local environmental contexts — to the issues, candidates and campaigning in any particular election. The analyses here have illustrated the impact of national issues which had locally-varying manifestations, such as the level of unemployment, and also of incumbents and campaign spending. Further, the residuals have identified places where specific changes have taken place, changes which may become deeply etched into the local political culture or which may disappear as rapidly as they appeared. Many of these alterations to the continuity reflect the arrival of the Alliance — the electoral base, fragile though it may be, provided by the SDP defectors, all but one of them from Labour; the by-election victories and near-successes; the ability to raise money to spend on campaigning. In some parts of England, the Alliance made substantial inroads, especially in the south where it appears to have replaced Labour as the potential opposition. Only future elections will show whether 1983 was critical and introduced new, enduring elements to the country's electoral geography, however.

It is common in analyses such as those presented here to suggest, at least implicitly, that general patterns of electoral behaviour are being uncovered. That approach has been resisted here, and the 1983 election results have been interpreted in their contemporary and historical contexts with no suggestion that the findings are transferable in time and space. They may be relevant to the future, but only the future will tell, because the geography of future election results will represent the interactions then of parties and voters, operating in historical contexts but seeking to overcome them. All that this book has shown is that the 1983 general election in England cannot be fully understood without a proper appreciation of the geography of electoral response to party agenda-setting.

This book is far from complete in its analyses of the geography of the 1983 results in England. As indicated in several places, more sophisticated analyses of the data sets are possible, evaluating alternative hyotheses to those presented here. Such analyses will undoubtedly not alter the basic conclusion outlined in the final sentence of the previous paragraph, although they might suggest slightly different interpretations. This book will have achieved its major purpose, however, if it ensures that no political scientist will in future criticise an analysis of electoral behaviour in England for failing 'to explore the relationship between geography, class and political affiliation' (Berrington, 1984, p. 119).

APPENDIX 1: THE MAP

One of the problems of mapping the geography of voting is the greatly varying areal extent of constituencies. As Kinnear's (1981) atlas shows, the high density of constituencies in the main urban areas can only be handled through the use of insets at larger scales; even so, the picture presented is dominated by the large constituencies in rural areas. (It is possible to show different parts of the country at different scales — as in Waller's 1983a, 1984b work — but this precludes a single, national map being used.)

A method of avoiding this problem is to transform the map so that each constituency occupies the same area, a procedure justified because each has approximately the same number of electors. Kinnear (1981, p. 12) disputes this, because 'constituencies have never had the same number of electors'. Since 1949, however, equality of electorate has been one of the criteria which the Boundary Commissions must apply when redefining constituencies. A new set of constituencies was introduced in February 1983. They were produced using 1976 data, however, and a court case sought to show that the equality criterion had not been sufficiently adhered to (Johnston, 1983f; Waller, 1983b). Nevertheless, in England only 17 constituencies deviated by more than 20 per cent from the national quota of 65,753, so it was decided that representing each constituency by a standard square was justified.

Kinnear (1981, p. 12) also criticises transformations into cartograms on the grounds that

> they do not accomplish what they set out to do. A map . . . is supposed to indicate where things happen.

This point has considerable validity, for it is indeed difficult to construct a cartogram in which the relative locations of all constituencies can be preserved. The main problems concern the major urban areas and conurbations. Nevertheless, it was decided to use a cartogram for the present book — although it was soon realised in the experiments at producing one that Greater London would have to be excluded from the main map. (The empty square in the centre of Greater London represents the City, a separate local government unit without separate Parliamentary representation.)

After many attempts, the cartogram used here was produced. The basic organising unit was the county, and as far as possible counties — and hence the constituencies within them — were placed in their correct relative locations. This proved impossible in the north of England, where Cumbria and Northumbria were of necessity separated — presumably by the uninhabited Pennines!

In the many maps in the text, the individual constituencies are not identified. They are on the map here (Figure A1.1). Further maps are included to identify some of the major towns (Figure A1.2) and the London boroughs (Figure A1.3).

The Constituencies in Numerical Order
The initials BC indicate Borough Council, according to the Boundary Commission description; CC indicate a county constituency.

1. Barrow and Furness CC
2. Berwick-upon-Tweed CC
3. Bishop Auckland CC
4. City of Durham CC
5. Copeland CC
6. Easington CC
7. Hexham CC
8. Langbaurgh CC
9. North Durham CC
10. North West Durham CC
11. Penrith and the Border CC
12. Sedgefield CC
13. Wansbeck CC
14. Westmorland and Lonsdale CC
15. Workington CC
16. Blaydon BC
17. Blyth Valley BC
18. Carlisle BC
19. Darlington BC
20. Gateshead East BC
21. Hartlepool BC
22. Houghton and Washington BC
23. Jarrow BC
24. Middlesbrough BC
25. Newcastle upon Tyne Central BC
26. Newcastle upon Tyne East BC
27. Newcastle upon Tyne North BC
28. Redcar BC
29. South Shields BC
30. Stockton North BC

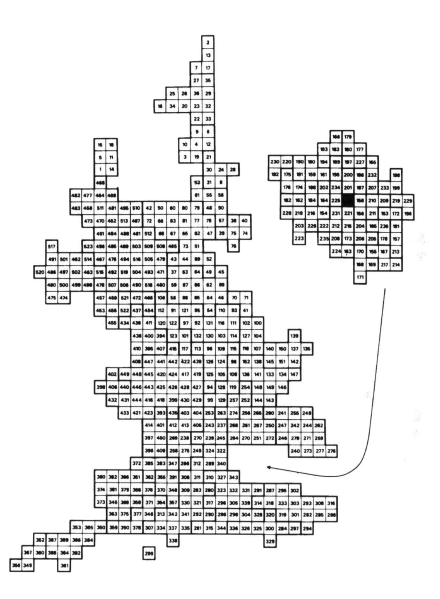

Figure A1.1 The 523 Constituencies on the Cartogram

31. Stockton South BC
32. Sunderland North BC
33. Sunderland South BC
34. Tyne Bridge BC
35. Tynemouth BC
36. Wallsend BC
37. Barnsley West and Penistone CC
38. Beverley CC
39. Boothferry CC
40. Bridlington CC
41. Brigg and Cleethorpes CC
42. Calder Valley CC
43. Colne Valley CC
44. Dewsbury CC
45. Doncaster North CC
46. Don Valley CC
47. Elmet CC
48. Harrogate CC
49. Hemsworth CC
50. Keighley CC
51. Normanton CC
52. Pontefract and Castleford CC
53. Richmond (Yorks.) CC
54. Rother Valley CC
55. Ryedale CC
56. Scarborough CC
57. Selby CC
58. Sheffield, Hallam CC
60. Shipley CC
61. Skipton and Ripon CC
62. Wentworth CC
63. Barnsley Central BC
64. Barnsley East BC
65. Batley and Spen BC
66. Bradford North BC
67. Bradford South BC
68. Bradford West BC
69. Doncaster Central BC
70. Glanford and Scunthro BC
71. Great Grimsby BC
72. Halifax BC
73. Huddersfield BC
74. Kingston upon Hull East BC
75. Kingston upon Hull North BC
76. Kingston upon Hull West BC
77. Leeds Central BC
78. Leeds East BC
79. Leeds North East BC
80. Leeds North West BC
81. Leeds West BC
82. Morley and Leeds South BC
83. Pudsey BC
84. Rotherham BC
85. Sheffield, Attercliffe BC
86. Sheffield, Brightside BC
87. Sheffield Central BC
88. Sheffield, Heeley BC
89. Wakefield BC
90. York BC
91. Amber Valley CC
92. Ashfield CC
93. Bassetlaw CC
94. Blaby CC
95. Bolsover CC
96. Bosworth CC
97. Broxtowe CC
98. Corby CC
99. Daventry CC
100. East Lindsey CC
101. Erewash CC
102. Gainsborough and Horncastle CC
103. Gedling CC
104. Grantham CC

105. Harborough CC
106. High Peak CC
107. Holland with Boston CC
108. Kettering CC
109. Loughborough CC
110. Mansfield CC
111. Newark CC
112. North East Derbyshire CC
113. North West Leicestershire CC
114. Rushcliffe CC
115. Rutland and Melton CC
116. Sherwood CC
117. South Derbyshire CC
118. Stamford and Spalding CC
119. Wellingborough CC
120. West Derbyshire CC
121. Chesterfield BC
122. Derby North BC
123. Derby South BC
124. Leicester East BC
125. Leicester South BC
126. Leicester West BC
127. Lincoln BC
128. Northampton North BC
129. Northampton South BC
130. Nottingham East BC
131. Nottingham North BC
132. Nottingham South BC
133. Bury St Edmunds CC
134. Central Suffolk CC
135. Great Yarmouth CC
136. Huntingdon CC
137. Mid Norfolk CC
138. North East Cambridgeshire CC
139. North Norfolk CC
140. North West Norfolk CC
141. South East Cambridgeshire CC
142. South Norfolk CC
143. South Suffolk CC
144. South West Cambridgeshire CC
145. South West Norfolk CC
146. Suffolk Coast CC
147. Waveney CC
148. Cambridge BC
149. Ipswich BC
150. Norwich North BC
151. Norwich South BC
152. Peterborough BC
153. Barking BC
154. Battersea BC
155. Beckenham BC
156. Bethnal Green and Stepney BC
157. Bexley Heath BC
158. Bow and Poplar BC
159. Brent East BC
160. Brent North BC
161. Brent South BC
162. Brentford and Isleworth BC
163. Carshalton and Wallington BC
164. Chelsea BC
165. Chingford BC
166. Chipping Barnet BC
167. Chislehurst BC
168. Croydon Central BC
169. Croydon North East BC
170. Croydon North West BC
171. Croydon South BC
172. Dagenham BC
173. Dulwich BC
174. Ealing, Acton BC
175. Ealing North BC
176. Ealing, Southall BC
177. Edmonton BC

178. Eltham BC
179. Enfield North BC
180. Enfield, Southgate DB
181. Erith and Crayford BC
182. Feltham and Heston BC
183. Finchley BC
184. Fulham BC
185. Greenwich BC
186. Hackney North and Stoke Newington BC
187. Hackney South and Shoreditch BC
188. Hammersmith BC
189. Hampstead and Highgate BC
190. Harrow East BC
191. Harrow West BC
192. Hayes and Harlington BC
193. Hendon North BC
194. Hendon South BC
195. Holborn and St Pancras BC
196. Hornchurch BC
197. Hornsey and Wood Green BC
198. Ilford North BC
199. Ilford South BC
200. Islington North BC
201. Islington South and Finsbury BC
202. Kensington BC
203. Kingston upon Thames BC
204. Lewisham, Deptford BC
205. Lewisham East BC
206. Lewisham West BC
207. Leyton BC
208. Mitcham and Morden BC
209. Newham North East BC
210. Newham North West BC
211. Newham South BC
212. Norwood BC
213. Old Bexley and Sidcup BC
214. Orpington BC
215. Peckham BC
216. Putney BC
217. Ravensbourne BC
218. Richmond and Barnes BC
219. Romford BC
220. Ruislip-Northwood BC
221. Southwark and Bermondsey BC
222. Streatham BC
223. Surbiton BC
224. Sutton and Cheam BC
225. The City of London and Westminster BC
226. Tooting BC
227. Tottenham BC
228. Twickenham BC
229. Upminster BC
230. Uxbridge BC
231. Vauxhall BC
232. Walthamstow BC
233. Wanstead and Woodford BC
234. Westminster North BC
235. Wimbledon BC
236. Woolwich BC
237. Aylesbury CC
238. Banbury CC
239. Beaconsfield CC
240. Billericay CC
241. Braintree CC
242. Brentwood and Ongar CC
243. Buckingham CC
244. Chelmsford CC
245. Chesham and Amersham CC

246. Epping Forest CC
247. Harlow CC
248. Harwich CC
249. Henley CC
250. Hertford and Stortford CC
251. Hertsmere CC
252. Mid Bedfordshire CC
253. Milton Keynes CC
254. North Bedfordshire CC
255. North Colchester CC
256. North Hertfordshire CC
257. North Luton CC
258. Oxford West and
     Abingdon CC
259. Rochford CC
260. Saffron Walden CC
261. St Albans CC
262. South Colchester and
     Maldon CC
263. South West Bedfordshire
     CC
264. South West Hertfordshire
     CC
265. Stevenage CC
266. Wantage CC
267. Welwyn and Hatfield CC
268. West Hertfordshire CC
269. Witney CC
270. Wycombe CC
271. Basildon BC
272. Broxbourne BC
273. Castle Point BC
274. Luton South BC
275. Oxford East BC
276. Southend East BC
277. Southend West BC
278. Thurrock BC
279. Watford BC
280. Aldershot CC
281. Arundel CC
282. Ashford CC

283. Basingstoke CC
284. Bexhill and Battle CC
285. Canterbury CC
286. Chichester CC
287. Dartford CC
288. Dover CC
289. East Berkshire CC
290. East Hampshire CC
291. East Surrey CC
292. Fareham CC
293. Faversham CC
294. Folkestone and Hythe CC
295. Gravesham CC
296. Guildford CC
297. Hastings and Rye CC
298. Horsham CC
299. Isle of Wight CC
300. Lewes CC
301. Maidstone CC
302. Medway CC
303. Mid Kent CC
304. Mid Sussex CC
305. Mole Valley CC
306. Newbury CC
307. New Forest CC
308. North Thanet CC
309. North West Hampshire
     CC
310. North West Surrey CC
311. Reading East CC
312. Reading West CC
313. Romsey and Waterside
     CC
314. Sevenoaks CC
315. Shoreham CC
316. South Thanet CC
317. South West Surrey CC
318. Tonbridge and Malling
     CC
319. Tunbridge Wells CC
320. Wealden CC

321. Winchester CC
322. Windsor and Maidenhead CC
323. Woking CC
324. Wokingham CC
325. Brighton, Kemptown BC
326. Brighton, Pavilion BC
327. Chertsey and Walton BC
328. Crawley BC
329. Eastbourne BC
330. Eastleigh BC
331. Epsom and Ewell BC
332. Esher BC
333. Gillingham BC
334. Gosport BC
335. Havant BC
336. Hove BC
337. Portsmouth North BC
338. Portsmouth South BC
339. Reigate BC
340. Slough BC
341. Southampton, Itchen BC
342. Southampton, Test BC
343. Spelthorne BC
344. Worthing BC
345. Bridgwater CC
346. Christchurch CC
347. Cirencester and Tewkesbury CC
348. Devizes CC
349. Falmouth and Camborne CC
350. Honiton CC
351. Northavon CC
352. North Cornwall CC
253. North Devon CC
354. North Dorset CC
355. North Wiltshire CC
356. St Ives CC
357. Salisbury CC
358. Somerton and Frome CC
359. South Dorset CC
360. South East Cornwall CC
361. South Hams CC
362. Stroud CC
363. Taunton CC
364. Teignbridge CC
365. Tiverton CC
366. Torridge and West Devon CC
367. Truro CC
368. Wansdyke CC
369. Wells CC
380. Westbury CC
371. West Dorset CC
372. West Gloucestershire CC
373. Weston-Super-Mare CC
374. Woodspring CC
375. Yeovil CC
376. Bath BC
377. Bournemouth East BC
378. Bournemouth West BC
379. Bristol East BC
380. Bristol North West BC
381. Bristol South BC
382. Bristol West BC
383. Cheltenham BC
384. Exeter BC
385. Gloucester BC
386. Kingswood BC
387. Plymouth, Devonport BC
388. Plymouth, Drake BC
389. Plymouth, Sutton BC
390. Poole BC
391. Swindon BC
392. Torbay BC
393. Bromsgrove CC
394. Burton CC
395. Cannock and Burntwood CC
396. Hereford CC
397. Leominster CC

398. Ludlow CC
399. Meriden CC
400. Mid Staffordshire CC
401. Mid Worcestershire CC
402. North Shropshire CC
403. North Warwickshire CC
404. Nuneaton CC
405. Rugby and Kenilworth CC
406. Shrewsbury and Atcham CC
407. South East Staffordshire CC
408. South Staffordshire CC
409. South Worcestershire CC
410. Stafford CC
411. Staffordshire Moorlands CC
412. Stratford-on-Avon CC
413. Warwick and Leamington CC
414. Wyre Forest CC
415. Aldridge-Brownhills BC
416. Birmingham, Edgbaston BC
417. Birmingham, Erdington BC
418. Birmingham, Hall Green BC
419. Birmingham, Hodge Hill BC
420. Birmingham, Ladywood BC
421. Birmingham, Northfield BC
422. Birmingham, Perry Barr BC
423. Birmingham, Selly Oak BC
424. Birmingham, Small Heath BC
425. Birmingham, Sparkbrook BC
426. Birmingham, Yardley BC
427. Coventry North East BC
428. Coventry North West BC
429. Coventry South East BC
430. Coventry South West BC
431. Dudley East BC
432. Dudley West BC
433. Halesowen and Stourbridge BC
434. Newcastle-under-Lyme BC
435. Solihull BC
436. Stoke-on-Trent Central BC
437. Stoke-on-Trent North BC
438. Stoke-on-Trent South BC
439. Sutton Coldfield BC
440. The Wrekin BC
441. Walsall North BC
442. Walsall South BC
443. Warley East BC
444. Warley West BC
445. West Bromwich East BC
446. West Bromwich West BC
447. Wolverhampton North East BC
448. Wolverhampton South East BC
449. Wolverhampton South West BC
450. Worcester BC
451. Bolton West CC
452. Chorley CC
453. City of Chester CC
454. Congleton CC
455. Crewe and Nantwich CC
456. Eddisbury CC
457. Ellesemere Port and Neston CC
458. Fylde CC
459. Halton CC

460. Hazel Grove CC
461. Heywood and Middleton CC
462. Knowsley North CC
463. Knowsley South CC
464. Lancaster CC
465. Littleborough and Saddleworth CC
466. Macclesfield CC
467. Makerfield CC
468. Morecambe and Lunesdale CC
469. Ribble Valley CC
470. South Ribble CC
471. Stalybridge and Hyde CC
472. Tatton CC
473. West Lancashire CC
474. Wirral South CC
475. Wirral West CC
476. Worsley CC
477. Wyre CC
478. Altrincham and Sale BC
479. Ashton under Lyne BC
480. Birkenhead BC
481. Blackburn BC
482. Blackpool North BC
483. Blackpool South BC
484. Bolton North East BC
485. Bolton South East BC
486. Bootle BC
487. Burnley BC
488. Bury North BC
489. Bury South BC
490. Cheadle BC
491. Crosby BC
492. Davyhulme BC
493. Denton and Reddish BC
494. Eccles BC
495. Hyndburn BC
496. Leigh BC
497. Liverpool, Broadgreen BC
498. Liverpool, Garston BC
499. Liverpool, Mossley Hill BC
500. Liverpool, Riverside BC
501. Liverpool, Walton BC
502. Liverpool, West Derby BC
503. Manchester, Blackley BC
504. Manchester Central BC
505. Manchester, Gorton BC
506. Manchester, Withington BC
507. Manchester, Wythenshawe BC
508. Oldham Central and Royton BC
509. Oldham West BC
510. Pendle BC
511. Preston BC
512. Rochdale BC
513. Rossendale and Darwen BC
514. St Helens North BC
515. St Helens South BC
516. Salford East BC
517. Southport BC
518. Stockport BC
519. Stretford BC
520. Wallasey BC
521. Warrington North BC
522. Warrington South BC
523. Wigan BC

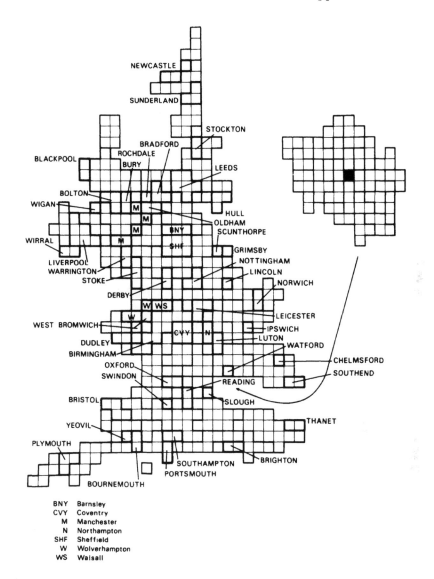

BNY Barnsley
CVY Coventry
M Manchester
N Northampton
SHF Sheffield
W Wolverhampton
WS Walsall

Figure A1.2 The Major Towns on the Cartogram

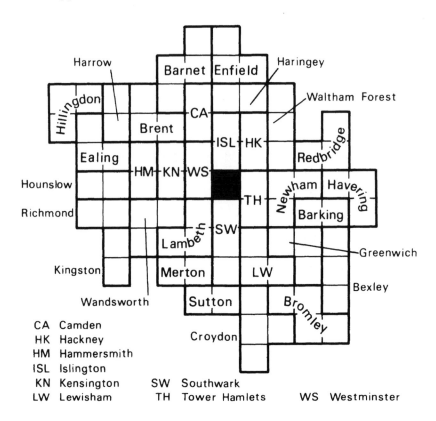

CA Camden
HK Hackney
HM Hammersmith
ISL Islington
KN Kensington   SW Southwark
LW Lewisham   TH Tower Hamlets   WS Westminster

Figure A1.3 The London Boroughs

APPENDIX 2: THE INDEPENDENT VARIABLES

The analyses in this book refer to a series of dependent variables, most of which have been mapped in the text. The independent variables have not been similarly displayed, however, so their geography is outlined here.

A. The Regional Division (Figure A2.1). The nine regions used here are those employed by OPCS. They comprise the following counties:

North — Cleveland, Cumbria, Durham, Northumberland, Tyne and Wear;

Yorkshire and Humberside — North, South and West Yorkshire, Humberside;

Northwest — Cheshire, Greater Manchester, Lancashire, Merseyside;

West Midlands — Hereford and Worcester, Shropshire, Staffordshire, Warwickshire, West Midlands;

East Midlands — Derbyshire, Leicestershire, Lincolnshire, Northamptonshire, Nottingham;

East Anglia — Cambridgeshire, Norfolk, Suffolk

Greater London;

Southeast — Bedfordshire, Berkshire, Buckinghamshire, East Sussex, Essex, Hampshire, Hertfordshire, Isle of Wight, Kent, Oxfordshire, Surrey, West Sussex;

Southwest — Avon, Cornwall, Devon, Dorset, Gloucestershire, Somerset, Wiltshire.

B. Constituency Type (Figure A2.2). This sixfold classification was used to produce an urban:suburban:rural classification, based on the counties and the division of constituencies into borough (relatively high density) and county (relatively low density). The six were

Metropolitan — borough;

Metropolitan — county;

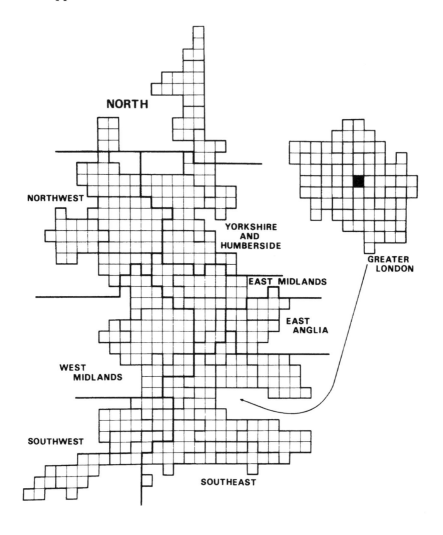

Figure A2.1 The Regional Division

Greater London — inner (the 12 Inner London Education
  Authority Boroughs — the pre-1963 London County
  Council);
Greater London — outer;
Nonmetropolitan — borough;
Nonmetropolitan — county.

C. Constituency Characteristics (Figures A2.3-A2.5). Five variables were selected to represent various aspects of the socio-

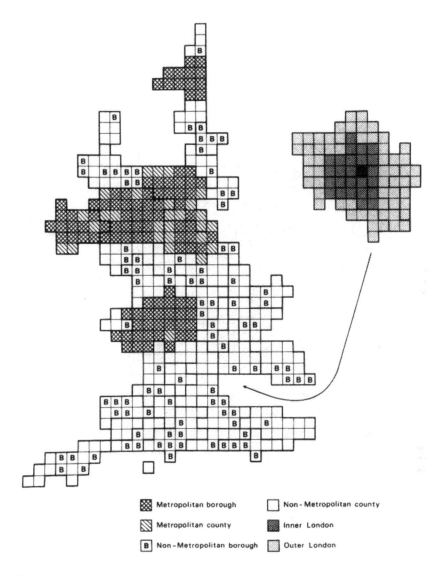

Metropolitan borough

Metropolitan county

B Non-Metropolitan borough

Non-Metropolitan county

Inner London

Outer London

Figure A2.2 The Constituency Types

economic characteristics of the constituencies. The first three referred to type of employment — in agriculture, energy (mining and power) and manufacturing; Figure A2.3 shows the constit-uencies with values more than one standard deviation above the mean for each of these. (Note that a few of the constituencies with above-average percentages in energy industries have no coal-mining but large power stations — e.g. Medway.) In this,

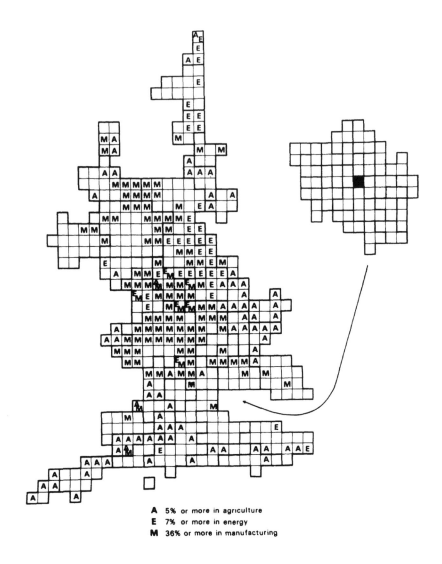

A   5% or more in agriculture
E   7% or more in energy
M   36% or more in manufacturing

Figure A2.3 Constituencies with Percentages More Than One Standard Deviation Above the Mean Employed in Each of Agriculture, Energy and Manufacturing

the concentration of agricultural employment in East Anglia and the Southwest is as would be expected, as is the concentration of employment in the energy industries in the Northeast and parts of Yorkshire and the East Midlands — though not in Lancashire. For manufacturing, the main concentration is in the Midlands

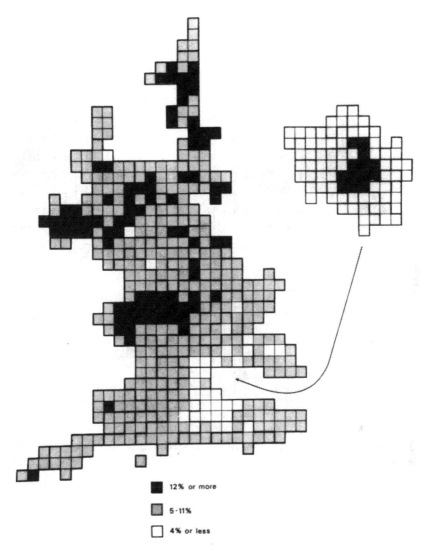

■	12% or more
▨	5-11%
□	4% or less

Figure A2.4 The Geography of Unemployment, 1981

and also in parts of the Home Counties north of London; the three northern regions have few constituencies with more than 36 per cent employed in manufacturing — indicative of the erosion of the industrial base there — whereas Greater London has none.

The geography of unemployment (the percentage of the economically-active population out of work) at the 1981 Census shows a very clear division of the country (Figure A2.4). In the

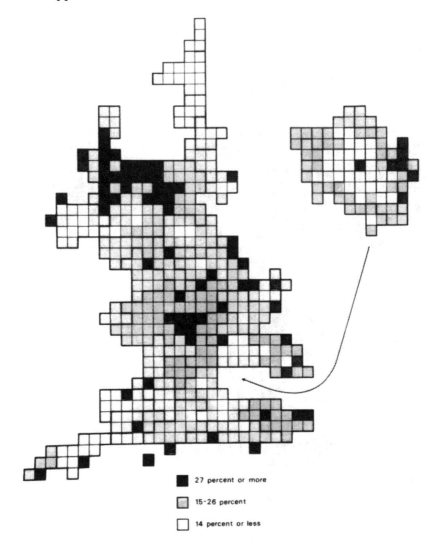

Figure A2.5 The Geography of Working-class Owner-occupancy

three northern regions only two constituencies — Cheadle and Ribble Valley — recorded a rate of four per cent or less, as did only one constituency in the East Midlands (West Derbyshire) and one in East Anglia (South East Cambridgeshire); in these regions, the major concentrations of unemployment of 12 per cent or more are in some of the main cities (Newcastle upon

Tyne, Sunderland, Stockton, Middlesbrough, Hull, Liverpool, Nottingham, Leicester, Birmingham, Coventry, Walsall, Warley, West Bromwich and Wolverhampton). Further south, there are only three pockets of unemployment more than one standard deviation above the national rate: two are single constituencies (Bristol South and Falmouth) whereas the other comprises a crescent of twelve inner London constituencies (in Newham, Tower Hamlets, Hackney, Lambeth and Southwark). The main concentration of constituencies with low unemployment rates forms another crescent, around the western boundary of Greater London.

The geography of working-class owner-occupancy (the percentage of working-class residents living in owner-occupier households; Figure A2.5) reflects two sets of factors — inter-regional 'cultural' variations and the spatially-varying impact of the 'right to buy' policy of the Conservative government; in 1981, the former was the most important. In the north of England there is a clear contrast between the low levels of owner-occupancy in Tyne and Wear and in Hull and the high levels in much of north Lancashire and neighbouring parts of Yorkshire (notably Bradford). The other concentration of working-class owner-occupancy is in Coventry. Greater London stands out with its well-below-average levels in 30 constituencies, mainly in Inner London; there are low levels, too, in most of the New Town constituencies (Basildon, East Berkshire, Crawley, Harlow, North Hertfordshire, Stevenage, Welwyn and West Hertfordshire).

The next three variables are the 'core classes' providing support for each of the main parties. For the Conservative Party, this was the percentage of the electorate in social class I or II, owner-occupied households (Figure A2.6). In this, the antici-pated north:south, urban:rural and inner city:suburban divisions of the country stand out, with the main bastion of potential Conservative support in a crescent of constituencies around the northern and western periphery of Greater London, including some Outer London suburbs, and along the south coast.

The core class for analyses of the Labour Party vote was the percentage of the electorate in working-class households. For this variable, the geography is, not surprisingly, virtually a mirror image of the previous one (Figure A2.7) — although there are slight differences, such as the absence of large concentrations of

18 percent or more

9–17 percent

8 percent or less

Figure A2.6 The Geography of Middle-class Owner-occupancy

working-class (i.e. manual occupation) households in east London.

Finally, the core class used in analyses of the Alliance vote was the percentage of the electorate in social class II households. Again, the geography (Figure A2.8) was very similar to that for the Conservative core class (the correlation between the two was

69 percent or more

45-68 percent

44 percent or less

Figure A2.7 The Geography of the Working Class

0.89; that between the Conservative and Labour core classes was −0.91 and between the Alliance and Labour cores −0.96). Some differences stand out, however, such as the concentration of potential Alliance support in North Yorkshire and the northern Home Counties.

Together, these three maps (Figures A2.6-A2.8) illustrate a

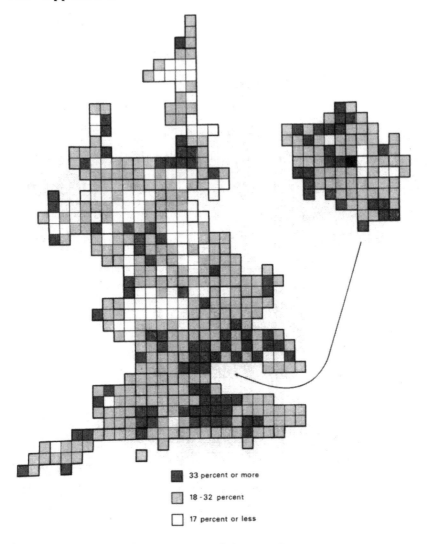

33 percent or more

18 - 32 percent

17 percent or less

Figure A2.8 The Geography of Social Class II

clear geographical separation in the bases of party support. Some places — virtually the whole of the North region, for example, the city of Hull, the South Yorkshire coalfield, the Merseyside district of Knowsley and the city of Sheffield (apart from the Hallam constituency, which has the highest percentage of the adult population with degrees in the whole country) — are very strongly pro-Labour; most are in the north, the West Midlands and London. Others are just as strongly pro-Conservative, and

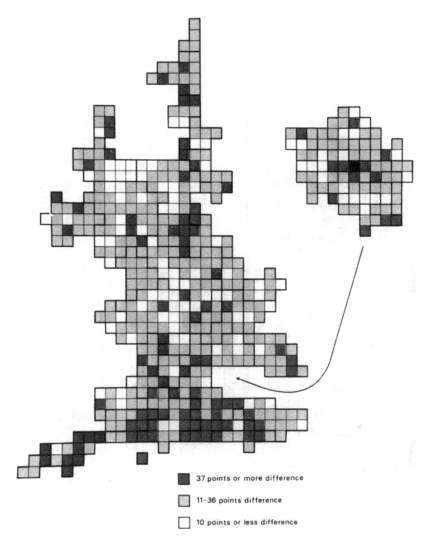

37 points or more difference

11-36 points difference

10 points or less difference

Figure A2.9 The Geography of Two-party Marginality

are mainly in the south; many of them also have the greatest potential support for the Alliance. This geography of social class is the foundation of the electoral geography of England, but, as the analyses here show, it does not provide a full explanation for the 1983 result.

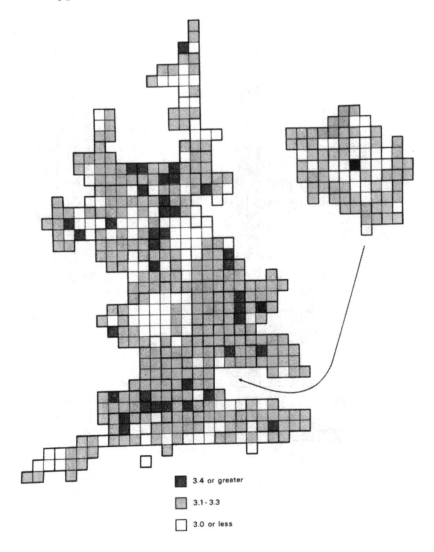

Figure A2.10 The Geography of Three-party Marginality

D. The Electoral Contest. Three sets of variables index aspects of the electoral contest in each constituency, referring to marginality, incumbency and the Alliance candidates.

The main index of marginality, the percentage points difference between Conservative and Labour in the 1979 estimated results, was chosen to cover the relative safety of seats in the

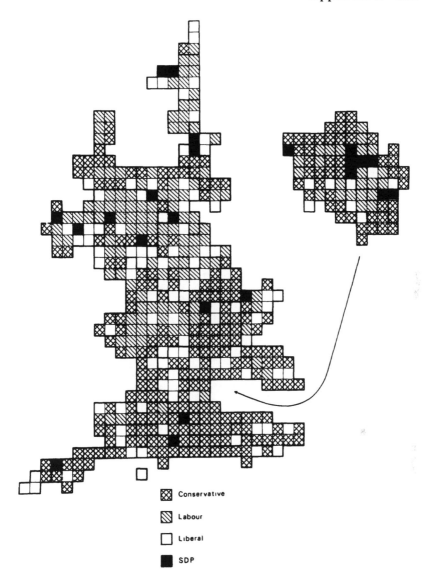

Figure A2.11 The Geography of Incumbents

great majority of constituencies, according to traditional partisan loyalties. The greater the index, the safer the position of one of the two parties. On this, the safest seats (those with an index of 37 percentage points or more) were in the south, especially the southern Home Counties and the south coast. In the North, most of the safest seats were in rural areas, with the exceptions of a

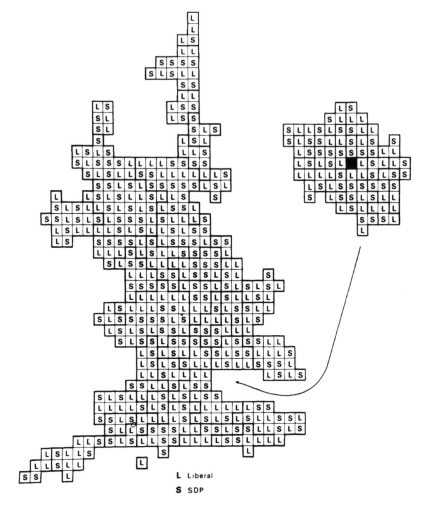

Figure A2.12 The Geography of Alliance Candidates

few mining constituencies and several in the Sheffield area (Figure A2.9). The Labour-Conservative marginals, on the other hand (those with a percentage points difference of 10 or less), were concentrated in urban areas — in North Lancashire and West Yorkshire, in the West Midlands, in the Hertfordshire New Towns, and in parts of Greater London.

A second marginality index was introduced in Chapter 9 to identify constituencies where all three parties had a 'winning chance'. The greater the H index, the closer the distribution of votes in the estimated 1979 results to 33 per cent for each party.

Relatively few constituencies had very high values (3.4 or more) on this index (Figure A2.10) — parts of East Anglia, Wessex and West Yorkshire had the main concentrations of seats in which all three parties were in a potentially winning situation. In much of the country, at least one party was weak and as a consequence the H value was relatively low — the Southwest, the West Midlands, Greater London, Sheffield/North Derbyshire, the Potteries and Tyne and Wear all comprise solid blocks of constituencies with index values of 3.0 or less.

Incumbency variables were introduced to inquire whether sitting MPs had a better chance of winning votes for their parties than did other candidates. The great majority of constituencies had an incumbent fighting the seat (Figure A2.11). Finally a variable was introduced to test for possible differences between the two Alliance Parties; the geography of the share-out is shown in Figure A2.12.

APPENDIX 3: THE ENTROPY-MAXIMISING PROCEDURE

The entropy-maximising procedure was introduced in Chapter 5 as a means of estimating the cell values in a three-dimensional matrix for which the sums — across rows, columns and faces — were known. The procedure is generally accepted as one which yields unique maximum likelihood estimates (i.e. those that are most likely to occur probabilistically: Bishop, Fienberg and Holland, 1975, ch. 3).

To illustrate this procedure (as in Johnston and Hay, 1982), take a matrix V with m rows and n columns, representing a national flow-of-the-vote matrix. This matrix $_mV_n$ can be subdivided into o constituencies, each of which has its own m × n matrix, $_mv_n$. Thus if

v_{ijk} is the value for the ith row, jth column and kth constituency then

$$\sum_{k=1}^{o} v_{ijk} = V_{ij}$$

The total number of voters in the country is V. The number of different combinations of voters in the first cell, v_{111} is then (given a fixed value of v_{111})

$$\frac{V!}{v_{111}! \, (V-v_{111})!}$$

For the next cell, v_{112} say, the number of combinations is

$$\frac{(V-v_{111})!}{v_{111}! \, (V-v_{111}-v_{112})!}$$

and so on for cell v_{113}, etc.

What is being done is that all the possible ways of allocating the V voters are being calculated, cell by cell, until all of the cells are filled. The total number of possible combinations is W, where

$$W = \frac{V!}{v_{111}! \, (V-v_{111})!} \times \frac{(V-v_{111})!}{v_{112}! \, (V-v_{111}-v_{112})!}$$

or

$$W = \frac{V!}{\pi \, v_{ijk}!}$$

which is the factorial of V divided by the product of the factorials of all v_{ijk} values. This can be represented, using Stirling's approximation, as

$$\log W = (V \log V - V) - \Sigma(v_{ijk} \log v_{ijk} - v_{ijk})$$

This expression can be evaluated for all possible combinations of v_{ijk}, so long as

$$\sum_{ijk} v_{ijk} = V$$

and the maximum-likelihood distribution of V voters through the v_{ijk} cells is the one that produces the largest value of W. Thus, if $m = 2$, $n = 2$ and $k = 4$ and $V = 32$, the value of W would be evaluated for every possible distribution of the 32 voters across the 16 cells.

In the evaluation, however, it is necessary to meet the constraints — the known values, which in a set of constituency flow-of-the-vote matrices are the two election results in each constituency and the national flow-of-the-vote. Thus, for example, in each constituency k

$$\sum_{i=1}^{m} v_{ijk} = v_{i.k}$$

where $v_{i.k}$ is the number of votes cast for party i (row i) at the first election, and

$$\sum_{j=1}^{n} v_{ijk} = v_{.jk}$$

where $v_{.jk}$ is the number of votes cast for party j (column j) at the second election. Over all constituencies

$$\sum_{k=1}^{o} v_{i.k} = V_{i.}$$

$$\sum_{k=1}^{o} v_{.jk} = V_{.j}$$

In other words, the sums of the row sums across all constituencies equal the relevant row sum in the V matrix, with the same being the case for the column sums. In addition

$$\sum_{k=1}^{o} v_{ijk} = V_{ij}$$

so that the sum of the values in any particular cell of the matrix, across all constituencies, is the relevant value in the V matrix.

These three constraints — row, cell and face — limit the distributions of V across all v_{ijk} to those that fulfil the conditions. This gives a statistic L, defined as

$$L = (V \log V - V) - (\sum_{ijk} v_{ijk} \log v_{ijk} - v_{ijk})$$

$$+ \sum_{ij} \lambda_{ij} (V_{ij} - v_{ij.}) + \sum_{ijk} \gamma_{ik} (V_{i.} - v_{i.k})$$

$$+ \sum_{jk} \phi_{jk} (V_{ij} - v_{ijk})$$

in which the parameters λ, γ and ϕ constrain the distribution of V to the known totals. The parameters can be calculated as

$$\exp(-\lambda_{ij}) = A_{ij}\, V_{ij}$$
$$\exp(-\gamma_{ik}) = B_{ik}\, V_{i.}$$
$$\exp(-\phi_{jk}) = C_{jk}\, V_{.j}$$

and the maximum likelihood values of v_{ijk} can then be calculated as

$$v_{ijk} = A_{ij}\, V_{ij}\, B_{ik}\, V_{i.}\, C_{jk}\, V_{.j}$$

This is done, as illustrated in Chapter 5, by an iterative procedure known as biproportionate matrix scaling (Bacharach, 1970). Initially, every value of v_{ijk} is set to 1.0. The sum of the values on one of the constraints is then compared to the known value, and a scaling factor derived which can be applied to the individual v_{ijk} values so that the two sums become equal (see Mosteller, 1968). The new values of v_{ijk} are then summed to another constraint and further scaling is undertaken. The scaling continues until the fit between all sums and constraints is within acceptable limits. Each individual value of v_{ijk} is then a unique combination of three scalars — that on the row constraint, that on the column constraint and that on the face constraint. There is a scalar for each row in each constituency (B_{ik}), for each column in each constituency (C_{jk}) and for each cell in the national flow matrix (A_{ij}). Each value of v_{ijk} — the most likely value for the flow i to j in constituency k — is the product of three constraints multiplied by the relevant scalars.

Entropy-Maximising and Log-Linear Models

This method of estimating the values of v_{ijk} is akin to the general procedure of log-linear analysis now widely used for the analysis of categorical data (see Wrigley, 1985; the following discussion follows Upton, 1978b). In this, the values of individual cells in a cross-classification are related to the constraints.

Take a 2×2 cross-classification matrix $_nV_m$, in which the row (i) and column (j) sums are

	20
	20

| 30 | 10 | |

This can be transformed into proportions of the total as

	0.5
	0.5

| 0.75 | 0.25 | |

These proportions can be interpreted as the probability of a randomly-selected individual being in the relevant row or column (i.e. the probability of being in row 1 is 0.5, etc.). Thus the probability of being in both row 1 and column 1 (cell v_{11}) is the product of the marginal probabilities (0.5 × 0.75), or

$$v_{ij} = (V_{i.}) \, (V_{.j})$$

where $V_{i.}$ is the probability of being in row i (i.e. it is the sum, across all j, of the probabilities in row i).

The previous paragraph suggests a very simple model, in which the value of a cell probability is a function of the interaction of the relevant two marginal probabilities. Log-linear analysis is based on models which estimate the influence of the marginal probabilities (the constraints) on known cell probabilities. Take the following table

	Vote (B)		
Class (A)	Conservative	Labour	
Middle	0.3	0.1	0.4
Working	0.2	0.4	0.6
	0.5	0.5	1.0

The goal is to estimate the influence of the constraint variables (class, A, and vote, B) on the cell values. The value of v_{11} (proportion of the population who are middle class and vote Conservative) is a function of: (1) the proportion of the population who are middle class (i.e. the larger the row total,

$V_{i.}$, the larger the value of v_{ij} is likely to be); (2) the proportion of the population who vote Conservative (i.e. the larger the column total, $V_{.j}$, the larger the value of v_{ij} is likely to be); and (3) the product of those two proportions (i.e. the larger the value of $(V_{i.})$ $(V_{.j})$ the larger the value of v_{ij} is likely to be). These three influences are modelled as follows

$$\log v_{ij} = \mu + \lambda_i^A + \lambda_j^B + \lambda_{ij}^{AB}$$

where

$\log v$ is the natural logarithm of the cell probability;

μ is the mean cell probability;

λ_i^A is the influence of the row constraint $(V_{i.})$

λ_j^B is the influence of the column constraint $(V_{.j})$

λ_{ij}^{AB} is the influence of the interaction of the two constraints

In this model since the parameters are expressing deviations around the mean (μ; equivalent to the constant term in a multiple regression equation) they must sum to zero; all the negative deviations from the mean must cancel out all the positive deviations. In a 2×2 classification this implies that

$$\lambda_1^A = -\lambda_2^A$$

$$\lambda_1^B = -\lambda_2^B$$

$$\lambda_{22}^{AB} = -\lambda_{12}^{AB} = -\lambda_{21}^{AB} = \lambda_{11}^{AB}$$

so that

$$\underset{i}{\overset{A}{\Sigma \lambda}}_{i} = \underset{j}{\overset{B}{\Sigma \lambda}}_{j} = \underset{j}{\overset{AB}{\Sigma \lambda}}_{ij} = \overset{AB}{\Sigma \lambda}{ij} - 0$$

The parameters of the model (the full derivation is in Upton, 1978b, p. 50) are

$$\lambda\overset{A}{_i} = V_{i.} - V_{..}$$

$$\lambda\overset{B}{_j} = V_{.j} - V_{..}$$

$$\lambda_{ij} = v_{ij} - V_{i.} - V_{.j} + V_{..}$$

and in natural logarithms these are (for examples in the 2 × 2 case)

$$\lambda\overset{A}{_1} = \frac{1}{4} \underset{j}{\Sigma} \log \left(\frac{v_{ij}}{v_{2j}}\right)$$

$$\lambda\overset{B}{_1} = \frac{1}{4} \underset{i}{\Sigma} \log \left(\frac{v_{il}}{v_{i2}}\right)$$

$$\lambda\overset{AB}{_{11}} = \frac{1}{4} \log \left(\frac{v_{11}\, v_{22}}{v_{12}\, v_{21}}\right)$$

which are the logarithms of the odds of being in a particular cell, given the row and column totals.

For the above cross-classification, where the natural logarithms of the cell values are

−1.204	−2.303
−1.609	−0.916

the values of the parameters are

$$\lambda_1^A = -0.2455 \qquad \lambda_2^A = 0.2455$$

$$\lambda_1^B = 0.1015 \qquad \lambda_2^B = -0.1015$$

$$\lambda_{11}^{AB} = 0.448 \qquad \lambda_{22}^{AB} = 0.448$$

$$\lambda_{12}^{AB} = -0.448 \qquad \lambda_{21}^{AB} = -0.448$$

and $\mu = -1.508$ (i.e. the natural logarithm of the mean cell probability, which is 0.25).
Thus

$$v_{11} = \mu + \lambda_1^A + \lambda_1^B + \lambda_{11}^{AB}$$

$$= -1.508 - 0.2455 + 0.1015 + 0.448 = -1.204$$

which shows that, relative to the mean probability, the row constraint has a negative influence on the proportion of the population who are middle class and voting Conservative (because the middle class are in a minority in the population), the column constraint has a positive influence and the interaction of the two constraints is positive (more middle-class people vote Conservative than one would expect from knowledge of how many middle-class people there are and how many Conservative voters there are).

The model can be applied to frequency data, rather than to probabilities, and can be expanded to a many-way classification. In the present context, where, for example, A may be the vote in 1979, B may be the vote in 1983 and C may be the national flow-of-the-vote, then

$$v_{ijk} = \mu + \lambda_i^A + \lambda_j^B + \lambda_k^C + \lambda_{ij}^{AB} + \lambda_{ik}^{AC} + \lambda_{jk}^{BC} + \lambda_{ijk}^{ABC}$$

so that the individual cell value is a function of the three constraints, operating independently of each other (the λ_i^A, λ_j^B and λ_k^C parameters), and of the interactions among all three. This procedure is used to model the influence of the constraints on known values of v_{ijk}. The entropy-maximising procedure produces estimates of v_{ijk}, given knowledge of the constraints. Those estimates are clearly derived from the most likely values of the parameters in the model. Thus, for example, an entropy-maximising estimate of the value of v_{ijk} in the equation in the previous paragraph incorporates estimates of each of the seven parameter values shown there.

Entropy-Maximising Estimates and Regression Analysis

In the procedure outlined here, the various constraints are used to evaluate the most-likely distribution of voters across the cells of a flow-of-the-vote matrix (or class-by-the-vote matrix) in each constituency. In the log-linear form of the estimating equation

$$v_{ijk} = A_{ij} \, V_{ij} \, B_{ik} \, V_{i.} \, C_{jk} \, V_{.j}$$

where $V_{i.}$ and $V_{.j}$ refer to the row and column totals.

The estimated values of v_{ijk} are used in the analyses here as the dependent variables. One of the hypotheses being tested is that v_{ijk} is a function of $V_{i.}$ (for example, that the number of Conservative voters remaining loyal to the party is a function of the number of Conservative votes cast in the constituency at the first election). Thus v_{ijk} is, in part, produced as a function of $V_{i.}$, according to the formula, and is then regressed against $V_{i.}$ to see if it is a function of the constraint. This would seem to be a tautology; the analysis is proving that which it assumes.

There is clearly some truth to this. However, as argued elsewhere (Johnston, Hay and Rumley, 1984), the value of v_{ijk} is a function of the product of three sets of constraints and their scalars. One of them — $V_{i.}$ in this case — may have little or no influence on the value of v_{ijk}, with the others, and the interactions among them, making the greatest contributions to the estimation. Thus the regressions reported here are not tautologous but are means of evaluating the relative importance of the row constraint (the 1979 election result for the analyses of the flow-of-the-vote matrices; the size of the relevant class for the analyses of the class-by-vote matrices) in the estimating proced-ure. If these were the only variables significantly related to the estimates, then the tautologous component would predominate. If, as is the case in all of the analyses, other variables are also significantly related to the estimates of v_{ijk}, then the geography of the dependent variables cannot be predicted solely by the geography of the constraints (i.e. Conservative loyalty is not just a function of 1979 Conservative vote). The unique value of v_{ijk} then reflects, at least in part, particular features of the constituency, which are explored by the regression analyses. The latter are attempts at decomposing the values of v_{ijk}, isolating the importance of B_{ik} and suggesting what may have influenced the values of A_{ij}, C_{jk} and their interactions. This argument can be rephrased in terms of the log-linear analysis discussed above. For a 2×2 classification, the model is

$$v_{ij} = \mu + \lambda_i^A + \lambda_j^B + \lambda_{ij}^{AB}$$

This is termed the fully-saturated model, in which all of the parameters are included and significant. It may be, however, that reduced forms of this model are sufficient. If, for example, all cell values are the same, then a sufficient model would be

$$v_{ijk} = \mu$$

Alternatively, it may be that only the row constraint influences deviations about the mean, so that

$$v_{ij} = \mu + \lambda_i^A$$

would be sufficient.

In the present context, the row constraint in the flow-of-the-vote analyses is the relevant 1979 vote, whereas in the class-by-vote analysis it is membership of the relevant class. If the model

$$v_{ijk} = \mu + \lambda \; {}_{i}^{A}$$

where λ_{i}^{A} is the parameter for the relevant row constraint, were sufficient then in the regression analyses with the values of v_{ijk} as the dependent variable only the independent variable relating to the row constraint would be significant; the tautology would be complete. If other independent variables were significant, however, this would indicate that other terms in the log-linear model were needed — perhaps all of them because the geography of voting was not a function of the one, inbuilt variable alone. This is what the present book demonstrates.

The procedure used in this book is based on widely-known methods of analysing categorical data. It produces maximum-likelihood estimates of the cell values in flow-of-the-vote matrices, at the constituency level. These estimates can then be analysed to identify their correlates, and so evaluate hypotheses regarding their geographical variations.

APPENDIX 4: SOME STATISTICAL ISSUES

The analyses in this book rely heavily on one family of statistical procedures — those based on the general linear model (Johnston, 1978). All of those analyses have been conducted using the well-known SPSS package of statistical routines especially the NEW REGRESSION routine in Release 7 (Nie *et al.*, 1975, 1983).

The use of these routines has not been very sophisticated, largely in the interests of ease of interpretation. There is plenty of scope for experimentation with other approaches — such as those based on path analysis. Nevertheless, a number of issues relevant to the procedures used are raised, and are outlined briefly here, justifying the decisions made.

The Use of Statistical Significance Tests

Throughout the book, statistical significance tests have been used in the evaluation of the various hypotheses; the criterion applied in all cases was that a statistic which was significant at the 0.05 level (in almost all cases, this was the F-statistic; see Nie *et al.*, 1983) was substantial enough to be interpreted. In all of the tables reporting regression equations, coefficients significant at that level have been asterisked.

Statistical significance tests were developed to allow inferences to be made about a defined population from study of a sample taken from that population. The strict interpretation of a statistically significant regression coefficient, therefore, is that the relationship which it describes (negative or positive) almost certainly occurs in the population — assuming a truly random sample. (Most tests relate the coefficient to a null hypothesis that the real value is zero, hence the statement about negative or positive: nothing about the size or strength of the relationship can be inferred unless the null hypothesis refers to an expected value other than zero.) Thus, according to this 'classical' view of inferential statistics, all that the results presented in this book indicate are relationships which almost certainly hold in the parent population.

What is the parent population? What are the 523 constituencies a sample of? To all intents and purposes, sample and population are the same, so that inference would appear to be irrelevant. Some workers argue that a study such as the present one is of a snapshot cross-section in time only, so that what we have is a sample of a large number of possible election results. This is not claimed here. There is no attempt to generalise away from 1983 and suggest that the findings here apply to other elections, past and future (although most of the hypotheses derive, of course, from the study of earlier elections). In any case, if this argument were carried to its logical conclusion then it could only be claimed that the sample size is *one*, not 523.

Another argument, particular to aggregate (or ecological) data analyses such as these, is that the set of constituencies used is but one of very many that might have been employed (i.e. might have been proposed by the Boundary Commission). Thus, it is argued, a statistically significant relationship is one that is likely to occur whatever the spatial aggregation of voters (again, therefore, a sample of one). However, work on the modifiable areal unit problem (see Openshaw, 1984) has shown that relationships can be very sensitive to the particular aggregation so that generalising from one may be very unwise. Further, some of the features of the individual constituencies — e.g. the incumbency variables — are not readily transferable. The 523 constituencies provide a singular set of observations, beyond which generalisation is impossible.

Why then use statistical significance tests? They are employed here merely as indicators of the strength of a relationship. The F value of a regression coefficient is an indicator of the amount of variation in the dependent variable that can be accounted for by the relevant independent — all other variables being held constant. It is, therefore, equivalent to the partial correlation coefficient for that relationship, and is an indicator of the degree of scatter about the regression line. The larger the value of F, the better the fit of the line. A significant value of F at the 0.05 level is thus used here as a convenient way of identifying a relatively strong relationship. It is preferred to the partial correlation since the latter gives no indication of the slope of the relationship, and it was felt unnecessary to report both — since the focus was on the slope (i.e. the nature of the relationship) and not on the relative importance of each independent variable. The value of F was not reported since the goal of the analyses was not to evaluate the relative importance of each independent variable; it was considered sufficient to report the 'firm' relationships and to focus attention on which these were. (In any case, because of the possibility of multicollinearity problems — see below — with substantial numbers of independent variables, such attempted evaluation could have been misleading.)

Closed Number Sets

Many aggregate data analyses transform the data for the areal units into percentages, proportions or other ratios, to remove the potential effects of a size variable. In the present case, that variable is either constituency electorate or constituency population in most cases. This practice can introduce problems for statistical analyses.

Some of these problems relate to what is known as the closed number set problem. If two or more variables with the same denominator with mutually exclusive numerators are employed, correlations may be in-built — i.e. they are a necessary part of the analysis and not a reflection of relationships between the variables. (See, for example, Johnston, 1978, and Evans and Jones, 1981.) For example, the two variables may be the percentage of a constituency's population aged over 45 and the percentage aged under 30; the numerators are mutually exclusive and the denominator the same. The larger the percentage aged over 45, the smaller the percentage aged under 30 is likely to be, especially if the former percentage is relatively large, since the two cannot sum to more than 100. (For an example, see Johnston, 1977.) As a consequence, there will almost certainly be a negative correlation between the two variables. This can influence the results of the statistical analyses and can introduce interpretation problems, because of multicollinearity (see below).

It is generally concluded that such closed number sets should be avoided, and only one percentaged variable from a mutually exclusive set based on the same denominator should be employed (Johnston, 1984a). In the present analyses, however, three such variables have been used:

(1) the percentage of the workforce employed in agriculture;
(2) the percentage of the workforce employed in energy industries; and

(3) the percentage of the workforce employed in manufacturing industries.

For these, the denominator was the industry of those in employment; a further independent variable, percentage unemployed, used a similar denominator — the economically-active population. (Three other variables — middle-class owner-occupiers; working class; social class II: see Appendix 2 — have the same denominator, but no more than one of them was used in the same analysis.) These three clearly qualify as part of a closed number set. However, two of them are very small in almost all constituencies, as the following summary statistics show:

	Agriculture	Energy	Manufacturing
Minimum	0.0	0.53	6.99
Maximum	18.56	33.57	53.48
Mean	2.05	2.97	27.40
Standard Deviation	3.05	4.05	9.12

The correlations between all pairs were low, too, as also were the correlations with the unemployment variable:

Agriculture	:	Energy	−0.06
Agriculture	:	Manufacturing	−0.23
Energy	:	Manufacturing	0.01
Agriculture	:	Unemployment	−0.29
Energy	:	Unemployment	0.02
Manufacturing	:	Unemployment	0.34

In no case was more than 12 per cent of the variation in one variable accountable by variation in the other, so it was concluded that all four variables could be included in a single regression equation without contaminating the results.

One other potential closed number set problem referred to three of the dummy variables. One of the regions was Greater London. Two of the types were Inner and Outer London. Within Greater London, clearly any constituency not in Inner London must be in Outer London, and vice versa. There is no correlation between the Inner and Outer London types, because only 82 of the 523 constituencies are involved in the two categories (the correlation is −0.09). There are substantial correlations, however, between the two types and the Greater London variable (0.58 and 0.71 respectively). This introduces potential multi-collinearity problems, with the possibility of large standard errors for the regression coefficients (and thus statistically insignificant findings). Experiments suggested that few potentially significant relationships were obscured by inclusion of all three variables; as many of the maps of the dependent variables and of the residuals show, there was considerable local variability in electoral behaviour within Greater London. For the sake of completeness, therefore, all three variables were included.

Multicollinearity

Collinearity arises in a multiple regression analysis when two of the independent variables are substantially correlated. The contribution of their shared variance to the 'explanation' of the dependent variable is obscured in the regression equation, and the partial regression coefficients, which may have large standard errors, refer only to the portion of the variation that is not shared. The larger the correlation between the pair of independent variables, therefore, the smaller the proportion of the total variation which the regression coefficients refer to (see Johnston, 1978). Multicollinearity occurs when three or more of the independent variables are substantially correlated, and the shared variation in even more diffuse in its distribution.

Closed number sets provide one source of multicollinearity, as discussed above. Other sources occur with independent variables that are correlated. The degree of multicollinearity can be identified by study of the correlations among the independent variables. Table A4.1 summarises these.

There are 24 independent variables which appeared in all of the regression analyses (the first four blocks in the table). Of the 276 different correlations involving pairs of variables from this set, only 16 exceeded $|0.04|$, and with very few exceeding $|0.5|$. (Apart from those involving the Greater London region and types, only three — the correlation between percentage unemployed and metropolitan borough; the correlation between unemployment and Conservative incumbent; and the correlation between Labour and Conservative incumbents — were greater than $|0.5|$.) This suggests a low level of multicollinearity: indeed 174 of the 276 correlation coefficients were less than $|0.2|$. (A principal components analysis of the common set of 24 inde-

Table   A4.1 Intercorrelations   Among   the   Independent Variables

	Largest Correlations		No. of Correlations	
	Positive	Negative	> |0.2|	> |0.4|
Agriculture	0.314	−0.288	8	0
Energy	0.284	−0.141	4	0
Manufacturing	0.373	−0.401	9	1
Working Class O-O	0.205	−0.479	8	3
Unemployment	0.528	−0.541	4	1
North	0.253	−0.234	2	0
Yorkshire/Humberside	0.335	−0.159	2	0
Northwest	0.386	−0.204	4	0
West Midlands	0.373	−0.174	2	0
East Midlands	0.221	−0.154	2	0
East Anglia	0.250	−0.104	1	0
Greater London	0.710	−0.401	7	3
Southeast	0.288	−0.362	5	0
Metropolitan Borough	0.528	−0.256	9	1
Inner London	0.573	−0.479	3	2
Outer London	0.710	−0.257	4	1
Metropolitan County	0.335	−0.109	2	0
Nonmetropolitan Borough	0.114	−0.232	3	0
Marginality	0.314	−0.395	4	0
Conservative Incumbent	0.312	−0.594	9	2
Labour Incumbent	0.453	−0.594	6	2
Liberal Incumbent	0.130	−0.136	0	0
SDP Incumbent	0.108	−0.217	2	0
Liberal Candidate	0.284	−0.217	2	0
Middle Class O-O	0.363	−0.723	11	5
Working Class	0.754	−0.642	8	4
Social Class II	0.649	−0.760	10	5
Conservative 1979	0.713	−0.618	13	3
Labour 1979	0.677	−0.664	12	6
Liberal 1979	0.524	−0.397	9	2
Conservative Spending	0.284	−0.416	7	1
Labour Spending	0.377	−0.664	10	1
Liberal Spending	0.216	−0.362	9	0

pendent variables confirmed this. The first component had an eigenvalue of only 3.63 — 15 per cent of the total variation — and there were ten components with eigenvalues exceeding 1.0.)

There were also two sets of independent variables, representing the 'core' classes and the 1979 election results, from which one was incorporated in nearly every regression analysis. Because only one of the six was included in a single equation, multicollinearity among the six was irrelevant, so only their correlations with the other 24 independent variables were identified for Table A4.1. About one-sixth of the correlations (24 of 144) exceeded |0.4|, although 44 per cent exceeded |0.2|. Most of the substantial correlations were those involving either a core class or a 1979 voting variable with one of: unemployment; incumbency; and marginality. There was then, some slight potential collinearity problem — which may, for example, account for the poor performance of the Conservative incumbency variable — but in general no difficulties were apparent and it was decided not to undertake any standardising procedures to try and circumvent the problem. The results suggested that this was not necessary.

Finally, there were the three spending variables, introduced in Chapter 9. The analyses there indicated the likelihood of some multicollinearity but in the event, as Table A4.1 indicates, it was relatively slight. The highest correlations were, as expected, with marginality; apart from these, there was nothing to suggest any potential problems.

The statistical analyses reported in this book were not very sophisticated, and relied almost entirely on the straightforward multiple regression model. The decision to depend largely on this one model was valid statistically, and few unreal assumptions had to be made. The result was a set of regression equations in which the coefficients were readily interpretable and the hypotheses easily tested. The correlates of the geography of voting were identified without difficulty.

The decision to adopt a relatively simple model, to promote ease of interpretation, meant that certain assumptions were made about the electoral behaviour analysed. It was assumed, for example, that the impact of the variables was linear (where there was any doubt, this was investigated and found to be justified) and additive; no interaction terms were included to see whether, for example, the relationships between 1979 vote and loyalty varied inter-regionally. More sophisticated models, such as those based on path analysis which can decompose the relationship between two variables into its direct and indirect elements, were ignored too. Again, this was largely for simplicity of presentation, given the general goal of the book.

There is, then, substantial scope for further, more sophisticated, analyses of these data. Given the success of the models tested, the results of such analyses are very unlikely to contradict the findings reported here. They may well amplify those findings, however, suggesting reasons for the small proportions of the geographical variation in voting in 1983 that remain unaccounted for.

APPENDIX 5: FLOW-OF-THE-VOTE MATRICES

Studies of changing patterns of voting behaviour which are not to rely entirely on inferences from cross-sections (i.e. the results at each election, and perhaps electoral surveys conducted then) require data on changing voter preferences. These can only be obtained from voter surveys, either panel surveys which interview the same voters at each election, and perhaps at intervening dates too, or cross-sectional surveys which ask people to recall how they voted in the past.

Both of these methods are fraught with problems, as Sarlvik and Crewe (1983) outline in their detailed Appendix on constructing such matrices. With panels, there are problems with respondents who drop out (i.e. decline to continue to provide data) and who lose contact with the survey (most because of migration). Others leave 'naturally' through either death or emigration (a very small proportion may be disenfranchised, because of imprisonment or certification as mad), and others join when they become enfranchised (the vast majority at age 18). The representativeness of the panel has to be maintained, therefore, as Butler and Stokes (1974) demonstrate in their detailed Appendix.

Although there are problems with maintaining panels, especially over long inter-electoral periods, the data yielded are almost always superior to those obtained from cross-sectional surveys, with recall questions to establish earlier voting behaviour. As Sarlvik and Crewe (1983, p. 346) indicate (see also Teer and Spence, 1973), in British surveys the usual problems of recall surveys are: (a) overestimates of Conservative and Labour voting in the past; (b) underestimates of Liberal and other (including Nationalist) voting in the past; (c) underestimates of past non-voting; and (d) underestimates of vote-switching (i.e. more people claim that they voted for the same party at the past and the present election than was actually the case). There are discrepancies between the recalled pattern of voting and the actual pattern, therefore, which require some standardisation of the data before detailed analysis.

Although panel data are preferable, nevertheless they are not without error, and Sarlvik and Crewe note that

> The data bases for estimating the flow of the vote between a pair of elections are fragile in the best of circumstances (p. 356)

They found it necessary to standardise the cell values of their estimated matrices, so that the sums of the various flows were the same as the actual election results. The procedure that they used — they call it Mostellerisation (p. 360) — involves exactly the same iterative matrix smoothing used in the entropy-maximising procedure employed here (Chapter 5 and Appendix 3: see Mosteller, 1968).

For the present study, no panel data were available and so a recall survey had to be relied upon. This was conducted on 9 June 1983 — election day — and so asked for recall of a voting preference expressed more than four years earlier (in May 1979). Not surprisingly, therefore, the matrix derived (Table 2.4) was not particularly accurate at 'postdicting' the 1979 election result. (To a small extent, this was due to the absence of any data on those who left the electorate between the two dates, although their impact was probably slight.) Thus the original matrix derived from the survey and without any reference to departures, had to be standardised using the entropy-maximising procedure, so that the estimated flows both postdicted the 1979 result, from the 1983 result, and predicted the 1983 result, from the 1979 result, within an average deviation of 5 per cent.

Clearly, the flow-of-the-vote matrices used here (and also the class-by-vote matrices) are coarse estimates. There is no reason to doubt their general validity, however (for they were used by one of the country's leading electoral analysts in his post-election articles: Crewe 1983a, 1983b).

The Changing Constituencies

Flow-of-the-vote matrices have been analysed by political scientists at the national level only. The goal of the present work was to investigate flows at the constituency level, which involved estimating a flow-of-the-vote matrix for each of the 523 constituencies, within the constraints of the 1979 results there, the 1983 result there and the national flow-of-the-vote matrix.

The 1979 constituencies were not used in 1983, however. A full review of all constituency boundaries was started in 1976 and completed in 1982 (Waller, 1983b; Johnston, 1983g), and was implemented in time for the 1983 election after a short delay caused by a legal challenge (Johnston, 1983f). There were 519 constituencies in England in 1979, only 60 of which were not altered at all in the review; a further 32 had changes affecting less than 5 per cent of the electorate. Thus, there was no set of 1979 results within which to set the study of changes.

Fortunately for this study, estimates of the voting in 1979 were produced for the new constituencies by teams of researchers working for the two television networks (BBC/ITN, 1983); the goal was to produce data against which the 1983 results could be compared in the election night programmes. This set of 1979 voting estimates was used in the analyses here.

Other analysts have been using the BBC/ITN estimates, and have suggested that they are deficient in some instances (for such analyses, see Steed and Curtice, 1983). They have identified 17 English constituencies where the 1979 estimates are dubious (I am grateful to John Curtice for providing me with this list), as follows:

(1) five constituencies in which the Conservative vote was underestimated and the Labour vote overestimated (Batley and Spen; Blackburn; Burnley; Don Valley; Ribble Valley);
(2) six constituencies in which the Conservative vote was overestimated and the Labour vote underestimated (Bolton Northeast; Brigg and Cleethorpes; Burnley; Calder Valley; Doncaster Central; Hyndburn);
(3) three constituencies in which the Conservative and Labour votes were underestimated and the Liberal vote overestimated (Hertsmere; Leeds Northwest; Manchester, Gorton);
(4) one constituency where the Conservative and Liberal votes were underestimated and the Labour vote overestimated (Denton and Reddish); and
(5) two constituencies where the Conservative and Labour votes were overestimated and the Liberal vote underestimated (Hertfordshire Southwest; Leeds West).

In most cases the number of votes redistributed by Curtice and Steed was small — less than 1,000 in several.

None of the constituencies in the above list stood out as a deviant case in the analyses of Chapters 7 and 8, so it can be assumed that the 'errors' in the BBC/ITN team's accounting did not substantially influence the results of the analyses undertaken here.

In order to undertake the analyses presented here, it has been necessary to use imperfect data — in the flow-of-the-vote matrices, the class-by-vote matrices and the estimated 1979 results. In some cases, approximations only have been used. Nevertheless, although such necessary 'short cuts' may have blurred the details of the picture slightly, they have not substantially distorted it. To present a valid geography of voting it has been necessary to deal in first approximations. The results amply justify the approach taken.

BIBLIOGRAPHY

Alford, R.R. (1963) *Party and Society*, Rand McNally, Chicago

Alker, H.R. (1969) 'A typology of ecological fallacies' in M. Dogan and S. Rokkan (eds.), *Quantitative Ecological Analysis in the Social Sciences*, The MIT Press, Cambridge, Mass., 69-86

Archer, J.C. and Taylor, P.J. (1981) *Section and Party*, John Wiley, Chichester

Bacharach, M. (1970) *Biproportional Matrices and Input-Output Changes*, The MIT Press, Cambridge, Mass.

Balsom, D. *et al.* (1983) 'The red and the green: patterns of partisan choice in Wales', *British Journal of Political Science*, 13, 299-325

BBC/ITN (1983) *The BBC/ITN Guide to the New Parliamentary Constituencies*, Parliamentary Research Services, Chichester

Berrington, H. (1983) 'The British general election of 1983', *Electoral Studies*, 2, 263-8

—— (1984) 'Decade of Dealignment', *Political Studies*, 32, 117-20

Bishop, Y.M.M., Fienberg, S.E. and Holland, P.W. (1975) *Discrete Multivariate Analysis*, The MIT Press, Cambridge, Mass.

Bodman, A.R. (1983) 'The neighbourhood effect: a test of the Butler-Stokes model', *British Journal of Political Science*, 13, 243-9

Bogdanor, V. (1983) *Multi-Party Politics and the Constitution*, The University Press, Cambridge

Bogue, D.J. and Bogue, E.J. (1982) 'Ecological correlation reexamined: a refutation of the ecological fallacy' in G.A. Theodorson (ed.) *Urban Patterns*, Pennsylvania State University Press, University Park, 88-103

Bonham, J. (1954) *The Middle-Class Vote*, Faber, London

Boucharenc, L. and Charlot, J. (1975) 'L'etude des transferts electoraux', *Revue francaise de science politique*, 24, 1205-18

Budge, I. and Fairlie, D.J. (1983) *Explaining and Predicting Elections*, George Allen and Unwin, London

Bulpitt, J. (1983) *Territory and Power in the United Kingdom*, Manchester University Press, Manchester

Burghardt, A.F. (1963) 'Regions and political party support in Burgenland', *The Canadian Geographer*, 7, 91-8

—— (1964) 'The bases of support for political parties in Burgenland', *Annals of the Association of American Geographers*, 54, 372-90

Butler, D.E. and Kavanagh, D. (1984) *The British General Election of 1983*, Macmillan, London

Butler, D.E. and Stokes, D. (1974) *Political Change in Britain* (2nd edn), Macmillan, London

Caldeira, G.A. and Patterson, S.C. (1982) 'Bringing home the votes: electoral outcomes in State legislative races', *Political Behavior*, 4, 33-67

Chilton, R. and Poet, R.R.W. (1973) 'An entropy-maximizing approach to the recovery of detailed migration patterns from aggregate census data', *Environment and Planning*, 5, 135-46

Clarke, P.F. (1971) *Lancashire and the New Liberalism*, The University Press, Cambridge

347

348 *Bibliography*

Converse, P.E. (1966) 'The concept of a normal vote' in A. Campbell *et al.* (eds.), *Elections and the Political Order*, John Wiley, New York, 9-39

Cox, K.R. (1969) 'The voting decision in a spatial context' in C. Board *et al.* (eds), *Progress in Geography, Volume 1*, Edward Arnold, London, 81-117

—— (1970a) 'Geography, social contexts, and voting behavior in Wales, 1861-1951' in E. Allardt and S. Rokkan (eds.), *Mass Politics*, The Free Press, New York, 117-59

—— (1970b) 'Residential relocation and political behavior: conceptual model and empirical tests', *Acta Sociologica*, 13, 40-53

Crewe, I. (1973) 'The politics of "affluent" and "traditional" workers in Britain: an aggregate analysis', *British Journal of Political Science*, 3, 29-52

—— (1976) 'Party identification theory and political change in Britain' in I. Budge *et al.* (eds.), *Party Identification and Beyond*, John Wiley, Chichester

—— (1981) 'Why the Conservatives won' in H.R. Penniman (ed.), *Britain at the Polls, 1979*, American Enterprise Institute, Washington, DC, 263-306

—— (1982) 'The Labour Party and the electorate' in D. Kavanagh (ed.), *The Politics of the Labour Party*, George Allen and Unwin, London, 9-49

—— (1983a) 'The disturbing truth behind Labour's rout', *The Guardian*, 13 June, 5

—— (1983b) How Labour was trounced all round', *The Guardian*, 14 June, 20

Crewe, I. and Payne, C. (1971) 'Analysing the Census data' in D.E. Butler and M. Pinto-Duschinsky, *The British General Election of 1970*, Macmillan, London, 416-36

—— (1976) 'Another game with nature: an ecological regression model of the British two-party vote ratio in 1970', *British Journal of Political Science*, 6, 43-81

Curtice, J. and Steed, M. (1982) 'Electoral choice and the production of government', *British Journal of Political Science*, 12, 249-98

Dicken, P. (1983) 'The industrial structure and the geography of manufacturing' in R.J. Johnston and J.C. Doornkamp (eds.), *The Changing Geography of the United Kingdom*, Methuen, London, 171-202

Draper, N.R. and Smith, H. (1966) *Applied Regression Analysis*, John Wiley, New York

Dunleavy, P. (1979) 'The urban basis of political alignment', *British Journal of Political Science*, 9, 409-43

—— (1980a) 'The political implications of sectoral cleavages and the growth of state employment. 1 The analysis of production cleavages', *Political Studies*, 28, 364-83

—— (1980b) 'The political implications of sectional cleavages and the growth of state employment: 2 Cleavage structures and political alignment', *Political Studies*, 28, 527-49

—— (1980c) *Urban Political Analysis*, Macmillan, London

—— (1982) 'How to decide that voters decide', *Politics*, 2 (2), 24-8

Dunleavy, P. and Husbands, C.T. (1984) *British Democracy at the Cross-roads: Voting and Party Competition in the 1980s*, George Allen and Unwin, London

Election Expenses (1983) Cnmd. 130, HMSO, London

Evans, I.S. and Jones, K. (1981) 'Ratios and closed number systems' in N. Wrigley and R.J. Bennett (eds.), *Quantitative Geography*, Routledge and Kegan Paul, London, 123-34

Fitton, M. (1973) 'Neighbourhood and voting: a sociometric explanation', *British Journal of Political Science*, 3, 445-72

Franklin, M.N. (1982) 'Demographic and political components in the decline of British class voting 1964-1979', *Electoral Studies*, 1, 195-220

Garrahan, P. (1977) 'Housing, the class milieu and middle-class conservatism',

British Journal of Political Science, 7, 125-6

Goldthorpe, J.H. *et al.* (1968) *The Affluent Worker: Political Attitudes and Behaviour*, The University Press, Cambridge

Gordon, I. and Whiteley, P. (1980) 'Comment. Johnston on campaign expenditure and the efficacy of advertising', *Political Studies*, 28, 293-4

Gudgin, G. and Taylor, P.J. (1979) *Seats, Votes and the Spatial Organisation of Elections*, Pion, London

Harris, R. (1984) 'Residential segregation and class formation in the capitalist city: a review and directions for research', *Progress in Human Geography*, 8, 26-49

Hawkes, A. (1969) 'An approach to the analysis of electoral swing', *Journal of the Royal Statistical Society A*, 132, 168-79

Himmelweit, H.T. *et al.* (1981) *How Voters Decide*, Academic Press, London

Hobson, J. (1968) 'Sociological interpretations of a general election' in P. Abrams (ed.), *The Origins of British Sociology, 1834-1914*, University of Chicago Press, Chicago, Illinois, 228-49

Houghton, Lord (1976) *Report of the Committee on Financial Aid to Political Parties*, Cmnd 6601, HMSO, London

Irwin, G. and Meeter, D. (1969) 'Building voter transition models from aggregate data', *Midwest Journal of Political Science*, 13, 545-66

Jacobson, G.C. (1980) *Money in Congressional Elections*, Yale University Press, New Haven, Conn.

Johnston, R.J. (1976) *The World Trade System*, G. Bell and Sons, London.

—— (1977) 'Principal components analysis and factor analysis in geographical research: some problems and issues', *South African Geographical Journal*, 59, 30-44

—— (1978) *Multivariate Statistical Analysis in Geography*, Longman, London

—— (1979a) 'Campaign expenditures and the efficacy of advertising at the 1974 general elections in England', *Political Studies*, 27, 114-19

—— (1979b) *Political, Electoral and Spatial Systems*, Oxford University Press, Oxford

—— (1980) *Geography of Federal Spending in the United States*, John Wiley, Chichester

—— (1981a) 'Regional variations in British voting trends 1966-1979: tests of an ecological model', *Regional Studies*, 15, 23-32

—— (1981b) 'Testing the Butler-Stokes model of a polarization effect around the national swing in partisan preference: England, 1979', *British Journal of Political Science*, 11, 113-17

—— (1981c) 'Embourgeoisement and voting in England, 1974', *Area*, 13, 345-51

—— (1981d) 'Short-term electoral change in England, 1974', *Geoforum*, 12, 237-55

—— (1982a) 'The changing geography of voting in the United States, 1946-1980', *Transactions, Institute of British Geographers*, NS7, 187-204

—— (1982b) 'The geography of electoral change: an illustration of an estimating procedure', *Geografiska Annaler*, 63B, 51-60

—— (1982c) 'Uncovering structural effects in ecological data: an entropy-maximising approach', *Geographical Analysis*, 14 355-65

—— (1982d) 'Embourgeoisement, the property-owning-democracy and ecological models of voting in England' *British Journal of Political Science*, 12, 499-503

—— (1983a) 'Spatial continuity and individual variability', *Electoral Studies*, 2, 53-68

—— (1983b) 'Class locations, consumption locations, and the geography of voting in England', *Social Science Research*, 12, 215-35

—— (1983c) 'The neighbourhood effect won't go away', *Geoforum*, 14, 161-8

—— (1983d) 'The feedback component of the pork barrel: tests using the results of the 1983 general election in Britain', *Environment and Planning A*, 15, 1567-76

—— (1983e) 'Campaign spending and voting in England: analysis of the efficacy of political advertising', *Government and Policy: Environment and Planning C*, 1, 117-26

—— (1983f) 'A reapportionment revolution that failed', *Political Geography Quarterly*, 2, 309-18

—— (1983g) 'Redistricting by independent commissions: a perspective from Britain', *Annals of the Association of American Geographers*, 72, 457-70

—— (1984a) 'Environmental influences and ecological analyses: examples from electoral geography', *Bremer Beitrage zur Geographie und Raumplanung*

—— (1984b) 'From Nixon to Carter: estimates of the geography of voting change, 1972-76', *Journal of Geography*, 82, 261-4

—— (1984c) 'Party strength, incumbency and campaign spending as influences on voting in four English general elections', *Tijdschrift voor Economische en Sociale Geografie*, 75

—— (1985) 'Peoples, places, parties and parliaments: a geographical perspective on electoral reform in Britain', *Geographical Journal*, 151

Johnston, R.J. and Doornkamp, J.C. (1983) 'Introduction' in R.J. Johnston and J.C. Doornkamp (eds.), *The Changing Geography of the United Kingdom*, Methuen, London, 1-12

Johnston, R.J. and Hay, A.M. (1982) 'On the parameters of uniform swing in single-member constituency electoral systems', *Environment and Planning A*, 14, 61-74

—— (1983) 'Voter transition probability estimates: an entropy-maximizing approach', *European Journal of Political Research*, 11, 93-8

—— (1984) 'The geography of ticket splitting: a preliminary study of the 1976 elections using entropy-maximising methods', *The Professional Geographer*, 36

Johnston, R.J., Hay, A.M. and Rumley, D. (1983) 'Entropy-maximizing methods for estimating voting data: a critical test', *Area*, 15, 35-41

—— (1984) 'On testing for structural effects in electoral geography, using entropy-maximizing methods to estimate voting patterns', *Environment and Planning A*, 16, 233-40

Johnston, R.J., Hay, A.M. and Taylor, P.J. (1982) 'Estimating the sources of spatial change in election results: a multi-proportionate matrix approach', *Environment and Planning A*, 14, 951-61

Johnston, R.J., O'Neill, A.B. and Taylor, P.J. (1983) 'The changing electoral geography of the Netherlands: 1946-1981', *Tijdschrift voor Economische en Sociale Geografie*, 74, 185-95

Johnston, R.J. and Semple, R.K. (1984) *Classification Using Information Statistics*, CATMOG 37, Geo Books, Norwich

Key, V.O. and Munger, F. (1959) 'Social determinism and electoral decision: the case of Indiana' in E. Burdick and A.J. Brodbeck (eds.), *American Voting Behavior*, The Free Press, Glencoe, Minnesota, 281-99

Kinnear, M. (1981) *The British Voter* (2nd edn), Batsford, London

Loosemore, J. and Hanby, V.J. (1971) 'The theoretical limits of maximum distortion: some analytic expressions for electoral systems', *British Journal of Political Science*, 1, 467-77

Luce, R.D. (1959) 'Analyzing the social process underlying group voting patterns' in E. Burdick and A.J. Brodbeck (eds.), *American Voting Behavior*, The Free Press, Glencoe, Il. 330-52

McAllister, and Rose, R. (1984) *The Nationwide Competition for Votes*, Francis

Pinter, London

McCarthy, C. and Ryan, T.M. (1976) 'Party loyalty at referenda and general elections: evidence from recent Irish contests', *Economic and Social Review*, 7, 279-88

—— (1977) 'Estimates of voter transition probabilities from the British general elections of 1974', *Journal of the Royal Statistical Society A*, 140, 78-85

McKenzie, R.T. and Silver, A. (1968) *Angels in Marble: Working Class Conservatives in Urban England*, Heinemann, London

McLean, I. (1973) 'The problem of proportionate swing', *Political Studies*, 21, 57-63

Madgwick, P. and Rose, R. (eds.) (1982) *The Territorial Dimension in United Kingdom Politics*, Macmillan, London

Miller, W.L. (1972) 'Measures of electoral change using aggregate data', *Journal of the Royal Statistical Society A*, 135, 122-42

—— (1977) *Electoral Dynamics*, Macmillan, London

—— (1978) 'Social class and party choice in England: a new analysis', *British Journal of Political Science*, 8, 257-84

—— (1979) 'Class, region and strata at the British general election of 1979', *Parliamentary Affairs*, 32, 376-82

—— (1981) *The End of British Politics? Scots and English Political Behaviour in the Seventies*, The Clarendon Press, Oxford

Miller, W.L., Raab, G. and Britto, K. (1974) 'Voting research and the population census 1918-1971: surrogate data for constituency analyses', *Journal of the Royal Statistical Study A*, 137, 384-411

Mosteller, F. (1968) 'Association and estimation in contingency tables', *Journal of the American Statistical Association* , 63, 1-28

Newby, W. (1977) *The Deferential Worker*, Allen Lane, London

Nie, N.H. *et al.* (1975) *SPSS*, McGraw-Hill, New York

—— (1983) *SPSS Update 7-9*, McGraw-Hill, New York

OPCS (1980) *Classification of Occupations, 1980*, HMSO, London

Openshaw, S. (1984) *The Modifiable Area Unit Problem*, CATMOG, 38, Geo Books, Norwich

Parker, A.J. (1982) 'The "friends and neighbours" voting effect in Galway West constituency', *Political Geography Quarterly*, 1, 243-62

Peake, L.J. (1984) 'Review essay: How Sarlvik and Crewe fail to explain the Conservative victory of 1979 and electoral trends in the 1970s', *Political Geography Quarterly*, 3, 161-8

Piepe, A., Prior, R. and Box, A. (1969) 'The location of the proletarian and deferential worker', *Sociology*, 3, 239-44

Pinto-Duschinsky, M. (1981a) *British Political Finance 1830-1980*, American Enterprise Institute, Washington, DC

—— (1981b) 'Financing the British general election of 1979' in H.R. Penniman (ed.), *Britain at the Polls, 1979*, American Enterprise Institute, Washington, DC, 241-62

Pool, I. de S. *et al.* (1960) *Candidates, Issues and Strategies*, The MIT Press, Cambridge, Mass.

Prescott, J.R.V. (1972) *Political Geography*, Methuen, London

—— (1978) 'Review', *Australian Geographical Studies*, 16, 9-11

Przeworski, A. and Soares, G.A.D. (1971) 'Theories in search of a curve: a contextual interpretation of the left vote', *American Political Science Review*, 65, 51-68

Pulzer, P. (1967) *Political Representation and Elections in Britain*, George Allen and Unwin, London

Putnam, R.D. (1966) 'Political attitudes and the local community', *American*

Political Science Review, 60, 640-54

Riecken, H.W. (1959) 'Primary groups and political party choice' in E. Burdick and A.J. Brodbeck (eds.), *American Voting Behavior*, The Free Press, Glencoe, Minnesota, 162-83

Robertson, D. (1976) *A Theory of Party Competition*, John Wiley, Chichester

Rokkan, S. (1970) *Citizens, Elections, Parties*, McKay, New York

Rose, R. (1967) *Influencing Voters*, Faber and Faber, London

—— (1974) 'Britain: simple abstractions and complex realities' in R. Rose (ed.), *Electoral Behavior*, The Free Press, New York, 481-541

—— (1976) *The Problem of Party Government*, Penguin, London

—— (1982a) 'From simple determinsim to interactive models of voting', *Comparative Political Studies*, 15, 145-69

—— (1982b) *Understanding the United Kingdom: The Territorial Dimension in Government*, Longman, London

Rumley, D. (1979) 'The study of structural effects in human geography', *Tijdschrift voor Economische en Sociale Geografie*, 70, 350-60

—— (1981) 'Spatial structural effects in voting behaviour: description and explanation', *Tijdschrift voor Economische en Sociale Geografie*, 72, 214-23

Runciman, W.G. (1966) *Relative Deprivation and Social Justice*, Penguin, London

Sarlvik, B. and Crewe, I. (1983) *Decade of Dealignment*, The University Press, Cambridge

Schattschneider, E.E. (1960) *The Semisovereign People: A Realist's View of Democracy in the United States*, Holt, Rinehart and Winston, New York

Segal, D.E. and Meyer, M.W. (1969) 'The social context of political partisanship' in M. Dogan and S. Rokkan (eds.), *Quantitative Ecological Analysis in the Social Sciences*, The MIT Press, Cambridge, Mass., 217-32

Senior, M.L. (1979) 'From gravity modelling to entropy maximizing: a pedagogic guide', *Progress in Human Geography*, 3, 179-210

Sharpe, L.J. (1982) 'The Labour Party and the geography of inequality: a puzzle' in D. Kavanagh (ed.), *The Politics of the Labour Party*, George Allen and Unwin, London, 135-70

SSRC Data Archive (1983) *SN1852: BBC 1983 Election Survey*, SSRC Data Archive, Colchester

Steed, M. and Curtice, J. (1983) *One in Four: An Examination of the Alliance Performance at Constituency Level in the 1983 General Election*, Association of Liberal Councillors, Hebden Bridge

Taylor, A.H. (1973) 'Variations in the relationship between class and voting in England, 1950 to 1970', *Tijdschrift voor Economische en Sociale Geografie*, 64, 164-8

Taylor, P.J. (1979) 'The changing geography of representation in Britain', *Area*, 11, 289-94

—— (1983) 'The changing political map' in R.J. Johnston and J.C. Doornkamp (eds.), *The Changing Geography of the United Kingdom*, Methuen, London, 275-90

—— (1984a) *Political Geography*, Longman, London

—— (1984b) 'Accumulation, legitimation and the electoral geographies within liberal democracy' in P.J. Taylor and J.W. House (eds.), *Political Geography: Recent Advances and Future Directions*, Croom Helm, London

Taylor, P.J. and Gudgin, G. (1982) 'Geography of elections' in N. Wrigley and R.J. Bennett (eds.), *Quantitative Geography: A British View*, Routledge and Kegan Paul, London, 382-6

Taylor, P.J. and Johnston, R.J. (1979) *Geography of Elections*, Penguin, London

Teer, F. and Spence, J.D. (1973) *Political Opinion Polls*, Hutchinson University

Library, London

Townsend, A.R. (1980) 'Unemployment and the new government's regional aid', *Area*, 12, 9-18

Upton, G.J.G. (1976) 'Diagrammatic representation of three-party contests', *Political Studies*, 24, 448-54

—— (1978a) 'A note on the estimation of voter transition probabilities', *Journal of the Royal Statistical Society A*, 141, 507-12

—— (1978b) *The Analysis of Cross-Tabulated Data*, John Wiley, Chichester

Waller, R. (1983a) *The Almanac of British Politics*, Croom Helm, London

—— (1983b) 'The 1983 Boundary Commission: policies and effects', *Electoral Studies*, 2, 195-206

—— (1984a) *The Dukeries Transformed*, Oxford University Press, Oxford

—— (1984b) *The Atlas of British Politics*, Croom Helm, London

Wanat, J. (1982a) 'The dynamics of presidential popularity shifts', *American Politics Quarterly*, 10, 181-96

—— (1982b) 'Most possible estimates and maximum likelihood estimates', *Sociological Methods and Research*, 10, 453-62

Wilson, A.G. (1970) *Entropy in Urban and Regional Modelling*, Pion, London

—— (1981) *Geography and the Environment: Systems Analytical, Methods*, John Wiley, Chichester

Wright, G.C. (1977) 'Contextual models of electoral behavior: the South Wallace vote', *American Political Science Review*, 71, 497-508

Wrigley, N. (1985) *Categorical Data Analysis in Geography*, Longman, London

INDEX OF PLACES

(Much of the analysis in this book focuses on the nine standard regions of England. These are excluded from the present index. Many of the places referred to are Parliamentary constituencies.)

SUBJECT INDEX

(Because the subject of this book is the general election of 1983, the three parties appear on virtually every page; they have not been included in this Index.)

357

Printed and bound by CPI Group (UK) Ltd, Croydon, CR0 4YY

22/10/2024

01777621-0016